TOWNLANDS

—

COUNTY TYRONE

WITH THEIR MEANINGS.

(SECOND IMPRESSION WITH CORRECTIONS, APRIL, 1936).
(THIRD IMPRESSION, 1988).

COMPILED BY

P. M'ALEER,

Some time Teacher of Irish under the Technical Department of Education.

This Edition Jan., 1987.
© **Ashardan** and **Moyola Books**
ISBN No. 0-9511836-1-3.

CONDITIONS OF SALE

This book is sold subject to the condition that it shall not, by ways of trade or otherwise, be lent, re-sold, hired out or otherwise circulated without the publisher's prior consent in any form of binding or cover other than that in which it is published and without a similar condition including this condition being imposed on the subsequent purchaser.

Published jointly by

MOYOLA BOOKS

Hill House, Owenreagh,
Draperstown, N. Ireland
BT 45 7BG.

and

Ashardan

12 Ballyoran Heights, Portadown,
Co. Armagh, N. Ireland.

PREFACE.

In the following pages I have endeavoured to give a reliable translation of all the townland names of Tyrone as contained in the Government Census record. I have been ably assisted by one (without whose aid I could not have undertaken the work) who, through the courtesy of the Keeper of the Records, Phoenix Park, Dublin, had access to the work of Dr. O'Donovan on Tyrone Place Names, and supplied me with a mass of useful information. I also consulted Dr. Joyce's well known "Names of Places," and venture to think the work may be looked upon as generally accurate and dependable.

My thanks are due to the above, and also to Messrs. Tobin, B.Sc., "Maboy," and W. Donnelly for many valuable suggestions and alternatives.

The undertaking, though arduous, was to me a labour of love, and I hope my efforts may be favourably received by the public.

P. M'ALEER.

FOREWORD

When I was approached by Mr. Graham Mawhinney and Mr. Eamon Jordan with the proposal that my father's book, "Townland Names of County Tyrone" (long out of print), should be re-published, I was delighted and assured them of my full co-operation.

I have had, over the years, requests from Tyrone emigrants in U.S.A., Australia, etc., for copies. Tyrone County Library, too, has had similar requests, so much so, that it was decided to have it microfilmed and as such can now be inspected at its head office in Omagh.

My father, until his death in 1941, was the Irish Folklore Commission's representative for the Sperrin district. After his death I carried out this work until the Commission sent a permanent representative in the person of Michael J. Murphy, playwright, writer and broadcaster, who recorded the vast amount of folklore and tradition for which this district is famous.

It was during my father's work and research in this field that he decided on the arduous task of the translation of "Townland Names". He discussed the project with my brother, Frank, M.A., who at the time was on the staff of St. Patrick's Teacher Training College Practising Schools, Drumcondra, of which he later became Principal. It was he, mentioned in the Preface, who helped with much of the research in Dublin.

On his retirement from teaching my father began his daunting task which entailed a visit to each separate townland, mostly carried out on a bicycle, noting the topographical and geographical features which form the basis of the gaelic name. He also concentrated on the local pronunciation of the townland names, which in many cases differed very much from their anglicised forms — so important when a translation was attempted. Mr. Glasgow of the Mid-Ulster Mail, who first published the book and later serialised it in his weekly paper, proposed a new edition to include a number of townland names inadvertently omitted in the first publication. He invited me to undertake this work but I declined as I was undertaking other work at the time. I very much regret the Post Office's decision to remove the townland as part of the address in rural districts. I trust that this book will save many townland names from oblivion.

I thank Mr. Mawhinney and Mr. Jordan for their efforts to retain some of our heritage.

Malachy C. McAleer,
Late Principal, Greencastle Primary School.
Christmas, 1987.

CASTLEDERG UNION.

PARISH—LONGFIELD WEST.

Aghakinmart (achadh cinn mart)—Hillhead of the cattle.
Annaghalough (eanach an locha)—Marsh of the lake.
Ally (aill)—A cliff or steep place.
Bomackatall (both Mhic a Tail)—M'Kattles hut.
Barravey (barr a bheithe)—Summit of the birch tree.
Billary (biolaire)—Abounding in water cress.
Carradoo (carra dubha)—Black weirs.
Carrick (carraig)—A rocky place.
Carrickbwee (carraig bhuidhe)—Yellow rock.
Clunahill (cluain eochaille)—Meadow of the yew trees.
Coolavannagh (cul a Mhanaigh)—The monk's corner (O'D.); or (cul da bheanach) corner of the two peaks; or (cul a bhaine) back of the milking place where cattle were milked (Joyce).
Cornashesk (cor na seisge)—Round hill of the sedge or coarse swampy grass.
Curragh Glebe (currach) A swampy place.
Collow (coll choill)—Hazel wood.
Curraghamulkin (currach a mhaoil chinn)—Moor of the bare hill.
Carrickaness (carraig an easa)—Rock of the cataract or waterfall.
Castlecraig (caislean craig)—Stone fort of the cliff.
Cavansallagh (cabhan salach)—Miry hollow.
Curraghmacall (currach Mhic Cathail)—Moor of Cathal's son.
Drumgallan (druim gallain)—Ridge of the pillar stone.

Drumnamalra (druim na malrach)—Ridge where youths congregate for amusement; or summit of the bartering or swopping.
Drumowen (druim Eoghain)—Owen's ridge; or (druim abhann)—The river ridge.
Drummenagh (druim meadhanach)—Middle ridge.
Drumquin (druim caoin)—Pleasant ridge; or (druim Cuinn) Conn's ridge.
Drumscra (druim scraithe)—Ridge of the sward.
Dunnaree (dun a fhraoich)—Fort of the heath or heather; or the King's fort.
Ednashanlaght (eadan an t-sean leacht)—Hillbrow of the old sepulchre.
Gortnasoal Glebe (gort na seal)—Loamy fields.
Kilmore (coill mhor)—Great wood or great church.
Kirlish (ciar lios)—Blackish fort.
Killan (coillin or cillin)—Little wood or church.
Killoan (coill Eoghain)—Owen's wood.
Lisky (lios sceiche)—Fort of the whitethorn bush.
Lackagh (leaca)—A hillside abounding in flat stones.
Longfield (leamhchoill)—The elm wood.
Meencargagh (min cairrgeach)—Rocky mountain field.
Marrock (mear-roc)—Brackish stream bed.
Meenadoan (min a duin)—Field or meadow of the fort.
Meenaheery (min na h-uidhre)—Mountain flat of the dun cow.
Meenmossogue (min measog)—The mountain meadow of the small red berries.
Meenacloy (min a chlaidhe)—Smooth meadow of the rampart or ditch.
Meenbog (min bog)—Soft mountain meadow.
Pruglish (pruchlais)—A cave.
Sloughan (leamhachan)—Place of elms, or a place of drooping foliage.
Tullyard (tulaigh ard)—High hill.
Tully (tulaigh)—A hill.

PARISH—TERMONAMONGAN.

Aghyaran (achadh Raitheachan)—A ferny field; or (achadh arann) the charioteer field.
Altamullan (alt a mullain)—Hillock of the deep glen side.
Aghamore (achadh mor)—Big field.

Aghascrebagh (achadh scriobach)— Furrowed or rugged field.
Aghnahoo (achadh na h-uaighe)—Field of the grave or cave.
Altgolan (alt gabhlan)—The forked glen.
Aghalougher (achadh locha)—Field of the lake or rushy field.
Aghalunny (achadh leamhnach)—Field of the elms; or (achadh loinne) of the gladness or gaiety.
Athabryanmore (achadh bruighne moire)—Field of the large fairy fort.
Ardarvar (ard arbhar)—Hill of the corn, or corn producing hill.
Ballymongan (baile Ui Mongain)—Mongan's town.
Clagernagh (clogerneach)—A place full of round hills or a place of much rain
Corgary (cor garbh)—Rough round hill.
Carrickaholten (carraig Ui Choltain)—Colton's rock.
Creeduff (crioch dubh)—Black boundary or mereland.
Crighdennis (crioch Denis)—Denis's boundary or district.
Crighshane (crioch Sheain)—John district.
Carndreen (carn draoighin)—Carn of the blackthorns.
Carnoughter (carn uachtair)—Upper carn or stone heap
Carracoghan (carn an chochain)—The carn of the straw or fodder.
Drummahon (druim meathain)—Ridge of the twigs or saplings.
Dreenan (draighean)—A place producing black-thorns.
Edenasop (eadan na sop)—Hillbrow of the wispy grass. or (t-sonna dhuin of the fortified dun.
Edenreagh (eadan riath)—Grey hill brow.
Essan (easan)—A miniature waterfall.
Golandun (gabhlan dun)—The dun or rath of the little fork.
Gortnacross (gort na croise)—Field of the cross.
Garvagh (garbh achadh)—Rough field or land.
Killen (coillean or cillin)—A little wood, or it might mean a little church.
Killeter (coill iochtair)—Lower wood or bottom land.
Legatonegan (lag a tonnaigain)—Pool of the wild duck or (t-sonna dhuin) of the fortified dun.
Lisnacloon (lios na cluaine)—Fort of the meadow.
Leitrim (liath druim)—Grey ridge.

Laghtfoggy (leacht fiaghaidhe)—The huntsman's monument.

Laghtmorris (leacht Muirgheasa)—Maurice's monument.

Lislaird (lios leirge)—Fort of the gentle hill.

Mourne Beg (mughdhorn beag)—Little town of the descendants of Mourne, son of Colla Meann.

Meenagrogan (min a ghruagain)—Field of the long grass.

Meenakeerin (min a chaorthain)—Field of the rowan trees.

Mullyfabeg (mulaigh fheadha big)—Summit of the little wood.

Mullyfamore (mulaigh an fheadha moir)—Summit of the big wood.

Meenacarriga (min na carraige)—Mountain field of the rock.

Meenclogher (min clocher)—Field of the stones.

Magherakeel (machaire chaol)—Narrow plain.

Meenafergus (min a bhFeargus)—Fergus's meadow.

Meenamullan (mine mullan)—The meadow of the round hill.

Magheranageeragh (machaire na g-caorach)—The plain of the sheep.

Mullanabreen (mullach na bruighin)—Summit of the small fairy fortress.

Sraghcumber (srath cumar)—The holm of the river confluence.

Scraghy (scraithe)—A sward or grassy surface of land.

Scralea (srath liath)—Grey holm.

Seegronan (suidhe Mhic Ronain)—Fairy mount of Ronan's son; or (gronndain) fairy hill of the lament or murmuring music.

Speerholme (spir holm)—A spit of land along a river side; or (speir thom) beautiful thicket or shrubbery

Tullycar (tul an choirr)—Rocky hill.

Tievenameenta (taobh na minte)—Side of the mountain meadow

Tulnashane (tullac na sion)—Stormy hill.

Trienamongan (trian na Mongain)—Mongan's third part.

Termonamongan (tearmann Ua Mongain)—Mongan's church land; or (na m-beangan) of the branches or boughs.

PARISH—KILSKEERY.

Brackagh (breacach)—Speckled land.
Bodoney (both Domhnaigh)—Sunday hut or prayer house.
Ballyard (baile ard)—High town.
Cabragh (cabrach)—Bad, rough, unprofitable land.
Coolback (cul baic)—Back of the bend; or the hollow-backed hill; or hill of the river-bend.
Cloncandra (cluain Chonraidh)—Conrath's meadow.
Corkill (cor coille)—The wooded round hill.
Corkragh (corcrach)—Purple coloured moory land.
Crossan (crosan)—A little cross.
Carran (carn)—A heap of stones; or
    (carran)—A place infested with spiraea or "yarr."
Cavanamara (cabhan na marbh)—Hill of the dead.
Cordromedy (cor droma fhada)—Hill of the long ridge; or (dromaide) a backband; hence cor dromaide, shaped like a straddle or backband.
Corlea (cor liath)—Grey hill or fort.
Doogary (dubh charra)—Black weir; also black surface; (dubh-chearradh) a large, or black, cutting or ravine.
Drumharvey (druim Harvey)—Harvey's ridge or hill.
Drumsonnus (druim sonnaigh)—Hill or ridge of the rampart or mound; or (druim sonais) prosperity ridge.
Derry (doire)—An oak grove.
Derryallen (doire alainn)—Beautiful oak grove.
Derrylea (doire liath)—Grey oak grove.
Dreigh (dreach)—Hill face or brae.
Drumash (drum ais)—Soft or wet ridge.
Drumdran (druim dreann)—Rough, scanty ridge.
Derrymacanna (doire M'Canna)—M'Cann's oak grove.
Drumbinnion (druim beinnin)—Little peak or spinked ridge.
Dernagilly (doire na gile)—Oak wood of brightness.
Drumardnagross (druim ard na gcros)—High ridge of the crosses; or it might mean high snout-shaped ridge.
Effernan (aifrionn)—A place where mass was formerly celebrated.
Feglish (fiodh glas)—Green wood.

Ferney (fearn mhaigh)—Plain of alders.
Garvaghy (garbh achadh)—Rough field or land.
Glasmullagh (glas mhullach)—Green summit.
Golan Glebe (gabhlan)—Fork-shaped land.
Gargaddis (garradh gadaidhe)—Garden frequented by thieves.
Greenan (grianan)—A sunny situation, land with a southern aspect.
Hackincon (achadh an chon)—Hound's habitation; also (achadh cinn con) field, etc., of warriors or hounds hill.
Killymendon (cille meantan)—Wood of snipes.
Kilknock (coill chnuic)—Hilly wood.
Killyblunick Glebe (coill na bluinige)—Wood of the lard.
Kilskeery (cill Scire)—St. Schirins wood.
Kinine (ceannin)—Small head of land; or (ceann eidhiun) ivy hill.
Knocknagor (cnoc na g-corr)—Hill of the cranes.
Keenogue (caonog)—Small mossy place.
Killyfuddy (coill fiude)—Cold or exposed wood; or (coill an fhuadaigh) perhaps wood of the plunder; or (coill na f-uaide) wood of the witch.
Lifford (liath bhear)—Grey waterside.
Lisdoo (lios dubh)—Black fort.
Lisnahanna (lios an eanaigh); fort of the marsh or cut-out bog; or (lios na h-ana) hill of the riches.
Loughterush (leacht a' ruis)—Monument of the wood or headland.
Makenny (maigh Cionnaith)—Kenny's plain.
Mulnagork (mul na g-corcach)—Summit of the marshes
Meeltogues (mioltoga)—A place of midges.
Magheralough (machaire locha)—Plain of the lake.
Moneygar (muine gearr)—Short shrubbery.
Relagh Guinness (raileach gineadha)—Productive land abounding in oak.
Realtons (reidh-altan)—Smooth hillock.
Roscor (ros corr)—Point or wood frequented by cranes.
Scallan (scalan)—A shed or a place of shadows.
Shanmullagh (sean mhullach)—Old hill top.
Screeby (scriobach)—Furrowed land.

Stralongford (srath long phuirt)—Holm of the fortress
Stranagummer (srath na g-cumar)—Holm of the meeting of the waters.
Tullywooly (tulach mhullaigh)—Hill of the peak or summit; or the peaked round hill; also (tulach bhuaile) the milking hill.
Tullynincrin (tulaigh n-aon crinn)—Summit of the solitary tree.
Trlllck (tri liaga)—Three remarkable standing stones.

CLOGHER UNION.

PARISH—AGHALURCHER.

Aghalurcher (achadh luachair)—Field of the rushes
Aghnaclogh (achadh na g-cloch)—Field of the stones or stony field.
Annagh (eanach)—A marshy place.
Artclea (ard cliath)—Heights upon which hurdles are got.
Ballagh (bealach)—A road or pass.
Beagh (beitheach)—A birchwood.
Ballymacan (baile Mhic Annaidh)—M'Cann's town.
Breakley (breach liath)—Speckled grey land.
Crookaclevin (cnoc a cliabhain)—Hill where baskets are made; or hill of the bare spot of grass.
Crocknahull (cnoc na h-olna)—Hill of the wool.
Cullontra (cuillean trach)—Holly-bearing land.
Cullynane (cuilleanan)—Holly land.
Fardross (far doras)—Lintel of a door.
Findermore (fionn rath mor)—Great white rath or fort
Killycorran (coill a cairn)—Wood of the stone heap or pile; or of the angle or corner.
Kilternan (cill Tighearnain)—St. Tiernan's church.
Lisboy (lios buidhe)—The yellow fort.
Loughermore (luachair mor)—Place of long rushes.
Mullaghmore (mullach mor)—Big summit.
Newry (uibhir)—A place abounding in yew trees.
Nurchossy (fuar choise)—Cold bottom lands.
Rehack (rath abhaic)—Level place of the dwarf.
       (rath acha)—Fort of the mound or rampart.
Relessy (reidh leasa)—Plain of the fort.
Slatbeg (sleaght beag)—The small monument or a place of rods or osiers.
Slatmore (sleaght mor)—The large monument.
Tullanavert (tulaigh na bhfeart)—Hill of the graves.
Tattanellen (taite Niallain)—Neyland's tate or division.
Timpony (tiompanach)—Round hillocks
Tircar (tir cairr)—Land (or district) of rocks.

PARISH—ERRIGAL KEEROGUE.

Altcloghfin (alt cloch fionn)—Glen or cliffs of white stones.
Altnagore (alt na n-gabhair)—Glen or cliff of the goats.
Annaghilla (eanach choille)—Marsh of the wood.

CLOGHER UNION. 13

Altamooskan (alt a muscain)—A perforated cliff covered with loose earth.
Ballygawley (baile Ui Dalaigh)—O'Daly's town; or (baile gaibhle) town or the forked roads.
Ballymackleroy (baile Mhic Giolla Ruaidh)—M'Elroy's town.
Brackagh (breacaidhe)—Speckled land, or a small plain or hill.
Ballylagan (baile an lagain)—Town of the hollow.
Ballynany (baile an eanaigh)—Town of the marsh.
Ballysaggart (baile sagairt)—Priest's town; could also mean town of the estate agent.
Coolageery (cul an ghaothraidh)—Hill back of the wood glen watered by a stream.
Cleanally (cluain aille)—Meadow of the slope or cliff.
Culnaha (cul na h-aithe)—Back of the lime kiln.
Carran (carn)—A heap of stones.
Cavey (cabhaigh)—A hollow or cavern.
Crew (craobh)—A large spreading tree.
Crossboy (cros bhuidhe)—Yellow cross.
Cullenbrone (cuillean bron)—Hollyland of the millstones; or hillback of grief or sorrow.
Derrymeen (doire min)—Pleasant oak wood.
Drumcorke (druim corcaighe)—Ridge of marshy ground.
Drumcullion (druim cuilinn)—Ridge of the holly trees.
Drumnamalta (druim na miolta)—Ridge of the cattle.
Dunmoyle (dun maol)—Bare fort.
Errigal Keerogue (aireagael)—Church or residence of St. Dachiarog.
Findrum (fionn druim)—White ridge.
Fernamenagh (fearann na manach)—Land of the monks or perhaps middle land.
Feddan (feadan)—A brook.
Foremass (for mas)—A round hill.
Fallaghearn (fal a fhearn)—Hedge of the alder trees; or (faill atha caorthainn) spink of the fort of the mountain ash.
Grange (grainseach)—Place for grain, generally attached to a monastery, a granary.
Garvaghy (garbh achadh)—A rough or uncultivated field.
Glenchuil (gleann coille)—Woody glen.

Gort (gort)—A field.
Killymorgan (coill Ui Morgain)—Morgan's wood.
Knockonny (cnoc na neona)—Hill of the cave, passage or souterrain.
Keady (ceide)—a green hill, level at top.
Kilgreen (coill greine)—Sunny woodland.
Knockbrack (cnoc breac)—Speckled hill.
Lettery (leitreach)—Hill of the spewy or wet side.
Lismore (lios mor)—Big fort.
Lisnabunny (lios na buine)—Fort of the springing or trickling of water.
Lisnawery (lios a bfoirreidh)—Fort of the level plain.
Lurganboy (lurganbuidhe)—Long yellow hill.
Roughan (ruadhachan)—Reddish coloured land.
Rarogan (rath ruagain)—Fort of the cold, dry breeze.
Shantavney (sean tamhnadh)—Old green field.
Sess (seiseadhach)—Sixth part of a townland.
Sess Kilgreen (seiseadhach coill greine)—Sixth part of the sunny wood.
Tirnaskea (tir na sgeach)—District of the briars or thorns.
Tullyglush (tulaigh ghlaise)—Hill of the luxuriant grass, a green hill.
Tullybryan (tulaigh Bhrian)—Brian's field or hill.
Tullylinton (tulaigh lanntain)—Hill of the commons, land not included in adjoining estates.

PARISH—ERRIGAL TROUGH.

Altadaven (alt an damhain)—Cliff side of the ox.
Carrickaoy (carraigan bhuaidh)—Rock of the victory.
Cullamore (cuaille mor)—Big pole or stake.
Dernasell (doire na seala)—Hill of the flocks.
Drumadarragh (druim na darach)—Ridge of the oak wood.
Durless (dur lios)—A strong fort.
Derrycloony (doire cluaineach)—Falling away oak wood.
Derryclay (doire cleithe)—Wood of the wattles.
Edenmore (eadan mor)—Big hillside.
Errigal Trough (aireagal Triucha)Church of the barony of Trough.
Fymore (flodh mor)—Big wood.
Gallagh (gallach)—Abounding in standing stones.
Killaveney (coill aoibhinn)—Beautiful wood.

PARISH—CLOGHER.

Altnaveigh (alt na bheagha)—Glen of the twigs, osiers and reeds.
Annaghloughran (eanach an lochain)—Marsh of the small lake.
Augher (eachradh)—An enclosure for cattle.
Annaghgarvey (eanach gairbhidhe)—Rugged, coarse marsh or road.
Ardunchin (ard uinseann)—Hill of the ash trees.
Annagh (eanach)—A marsh.
Aghamilkin (achadh minchoille)—Field of the smooth or bare wood.
Aghindarragh (achadh na darach)—Field of the oak.
Aghandrumman (achadh dromainn)—Field of the big ridge.
Altnaveragh (alt na bhfear-mhaigh)—Cliff of the grassy plain.
Aghintain (achadh an tamhain)—Field of the tree stump.
Aghnagloch (achadh na g-cloch)—Stony field.
Aghangowley (achadh na gabhlaidhe)—Field of the forked road or river, etc.
Aghinlark (achadh an leirg)—Field of the hillside.
Aghnacraney (achadh na gcairne)—Field of the cairns or stone heaps.
Ballywholan (baile a bhullain)—Town of the judge or of the round hollow.
Ballaghneed (bealach an ide)—Pass of the massacre.
Ballynese (baile an easa)—Town of the waterfall.
Beigh (beitheach)—Birch land.
Bolies (bualaidhe)—A dairy or milking place.
Ballagh (bealach)—A road or pass.
Ballymacan (baile Mhic Annaidh)—M'Cann's town.
Ballyvadden (baile Ui Mhadadhain)—Maddenstown.
Ballygrunan (baile grianain)—Town of the sunny aspect; a palace.
Ballynagurragh (baile na g-currach)—Town of the moors or soft marshes.
Branter (brean tir)—Town of the bad smells, or possibly (brainteoir) a quern worker.
Belnaclogh (beul na g-cloch)—Ford mouth of the stones.
Beltany (bealtaine)—May day or town of the bonfires.

Ballynagowan (baile na ngabhann)—Blacksmiths' town

Ballyscally (baile scailtigh)—House with open front, used as a mass house in penal times.

Cargagh (cairrgeach)—Rocky lands.

Carrickavoy (carraiga bhuaidhe)—Rock of the victory.

Cavan (cabhan)—A hill; the word also means a hollow or low-lying place.

Cloonycoppoge (cluain na g-copog)—Meadow of the dock leaves.

Corcloghy (cor clochach)—Round stony hill.

Carntall Beg (carn Tal beag)—Tal's small carn or stone heap; Tal is a person's name.

Carntall More (carn Tal mor)—Tal's big carn.

Corboe (cor both)—Round hill of the hut.

Corkhill (corr choille)—Round or corner hill of the wood.

Cavanacark (cabhan na gcurc)—Hollow of the bushy tufts.

Carnagat (carn na g-cat)—The cat's rock.

Cole (cul)—Back of a hill.

Corcreevy (cor craoibhidhe)—Bushy hill.

Caldrum (coll druim)—Ridge of the hazels.

Carr (cor)—A hill.

Cornamucklagh (cor na muclach)—Hill of the piggeries

Clare More (clar mor)—The big plain.

Cloneblaugh (cluain blathach)—Flowery meadow.

Cormore (cor mor)—Great round hill.

Carnahinny (carn na teine)—Rock on which fires were lighted; or carn of the whins.

Carryclogher (caraidh clochair)—Weir of the carry; or (cathair a clochair)—The church of the convent.

Cloughlin (cloch liathan)—The grey stone.

Clogher (cloch oir)—Golden stone. This stone was worshipped under the name of Kermann Kelstack and said to be preserved in Clogher Church. More likely to mean a stony place. See Joyce, Vol. 1, p. 413.

Corick (comh rac)—A confluence or meeting place of rivers or streams.

Corleaghan (cor leaghan)—Wide or broad hill.

Crossowen (cros Eoghain)—Owen's cross.

Cargagh (cairgeach)—Rocky lands.

Derries (doiridhe)—Oak woods.

Derrynascobe (doire na scuab)—Wood where brooms are obtained; or (doire na scuaib gaoth) the oak wood of the whirlwind.

Divinagh (dubh eanach)—Black marsh or cut out bog.

Dunbiggan (dun buigiuin)—Fort of the rank or soft growths.

Drumhirk (druim thuirc)—Hog's ridge.

Derrydrummond (doire dromain)—Oak wood of the long hill.

Dromore (druim mor)—The big ridge.

Donaghmoyne (domhnach maighin)—Church of the little plain.

Eskermore (eiscer mor)—Great gravelly sand hills.

Eskernabrogue (eiscer na m-brog)—Shoe shaped eskers.

Eskragh (eiscreach)—Abounding in ridges of sand hills

Edergole (eadar gabhla)—Place between fork shaped projections of land.

Farranetra (fearann iochtrach)—Lower land.

Fernaghandrum (fearnan an droma)—Land of the ridge.

Freughmore (fraoch mor)—Great heath.

Fogart (fo gart)—Good productive land.

PARISH—CLOGHER.

Findermore (fionn rath mor)—Great white rath or fort; or the big, beautiful hill.

Fardross (far dris)—Great brambly or briery spot.

Glenhoy (gleann thuaidh)—North valley or glen.

Glennageeragh (gleann na g-caorach)—Glen of the sheep.

Grange (grainseach)—A granary.

Gunnell (g-coinneal)—A place of fairy lights.

Garlaw (garbhlach)—A rough place or rough road.

Glennoo (glen uagha)—Glen of the cave.

Gortmore (gort mor)—A big field.

Kilclay (coill cleithe)—Mountain top wood; or (neamh eirightheach) unprosperous wood or church.

Kilnaheery (coill na h-uidhre)—Wood of the brown or dun cow

Knockmany (cnoc manaigh (O'D.)—The monk's hill, or more likely (cnoc meadhonach) middle hill.

Kilruddan (coill Rodain)—Rodden's wood.

Killaney (coill Eanaidh)—Eany's or Ethna's wood; or (coill leane) the wood meadow.
Killyfaddy (coill phaiteach)—A humpy wood.
Kilnahushogue (coill na h-uiseoige)—The wood of the lark.
Knocknacarney (cnoc na ceithirne)—The Kern's hill or hill of the band of warriors.
Killycorran (coill a chairn)—Wood of the stony heap.
Killygordon (coill an ghabhair duinn) wood of the brown goat.
Latbeg (leacht beag)—A small monument.
Lisnarable (lios na riobal)—Fort of the long stones.
Lismore (lios mor)—Big fort.
Lisnamaghery (lios na machaire)—Fort of the plain.
Lisbane (lios ban)—White fort.
Lisgorran (lios gorain)—Fort of the oozey or wet soil
Lislea (lios liath)—Grey fort.
Lurganaglare (lurga na g-clar)—A long plain or hill.
Lislane (lios leaghan)—Broad or wide fort.
Lungs (longa)—Site where many houses were built.
Lisboy (lios buidhe)—The yellow fort.
Losset (lusaid)—A k neading trough or losset or place of herbs.
Mullaghmore (mullach mor)—Big summit.
Mullans (mullain)—Little hills.
Mullaghtinny (mullach teine)—Summit upon which fires used to be lighted; or (mullach teimhne) the hill of darkness.
Mallabeny (malaidhe beannai)—Braes of the spinks.
Nurchossy (fhuar coise)—Cold bottom land of the cool stream.
Prolusk (prochlais)—A cave or badger den.
Roy (rath)—A fort or dun
Ratory (rath toraidhe)—Parched or dry fort.
Rahoran (rath an bhurain)—Fort of the spring well.
Rosemeilan (ros Maolain)—Moylan's fort.
Ranenly (rath Niale)—Neill's fort.
Sessia (seiseadhach)—A sixth part or division.
Shanoo (sean chaoi)—Old road.
Skelgagh (sgeilgeach)—Rocky ground.
Shantonagh (sean tamhnach)—Old green field
Syunchin (suidhe fhuinnseog)—Abounding in ash trees.

Slatbeg (slat beag)—A small wood abounding in rods or osiers.

Slatmore (slat mor)—A large wood abounding in rods or osiers.

Screeby (scriobach)—Furrowed or rugged land.

Sess (seiseadhach)—A sixth part or division.

Tully (tulaigh)—A hill.

Tamlaght (tamhlacht)—A burial place or plague monument.

Tullanafoile (tullach na f-aille)—Hill of the cliff; or hill of the cavern.

Tullycorker (tulaigh chorcair)—Purple coloured hill.

Tullyvernan (tulaigh bhearnain)—Hill of the little gap.

Tycanny (tigh ceannaidhe)—The merchant's house or shop.

Terrew (triuch)—A district or country side.
 (I don't see the connection, but cannot give any other.)

Tullybroom (tullach bhruim)—Morose or sullen-looking hill.

Tawnymore (tamhnach mor)—Big green field.

Tatnadaveny (tate damhan)—Division of the oxen; or (na doimhne) division of the deep soil.

Townagh (tamhnach)—A green field.

Tullyquinn (tulaigh Chuinn)—Quinn's hill; or (tulagh con) quiet clean hill.

Tullanavert (tulaig na bhfeart)—Hill of the graves.

Tattanafinnell (tate na fionghaile)—Field of the fratracide.

COOKSTOWN UNION.

PARISH—ARDBOE.

Ardboe (ard both)—The high hut.
Annaghmore (eanach mor)—Great marsh.
Aneeter Beg (an iocthar beg)—Little low lying ground
Aneeter Mor (an iochtar mor)—Big low lying ground.
Ardean (ardan)—Little height or hillock.
Aghacolumb (achadh Colum)—Columb or Coleman's field.
Ballynargin (baile n-argain)—Town of the plunderings.
Ballynafeagh (baile na faithche)—Town of the green plots or wattles.
Ballymaguire (baile Mhic Uidhir)—M'Guire's town.
Ballymurphy (baile Murchadha)—Murphy's town.
Cluntoe (cluainte)—Meadowy land.
Carnan (carnan)—Little carn or heap of stones.
Drumenny (druim eanaigh)—Deer backed summit, or perhaps watery ridge.
Drumard (druim ard)—High ridge.
Drumhubbert (druim tubrid)—Ridge of the "tubrid" or well.
Dromore (druim mor)—Big ridge.
Eary (airghe or aireadh)—A cultivated division of land or pasture.
Elagh (aileach)—A stone house or fort.
Farsnagh (Fairsingeacht)—A wide roomy place.
Feagh (foidhach)—Woody arbutus land.
Gortigal (gort geal)—Bright field.
Gortnawyg (gort na diobhoige)—Field of the rivulet running through boggy land.
Killycanavan (coill an cheannbhain)—Wood of the wild or bog cotton.
Killygonlan (coill Ui Gonnlain)—O'Gonnalan's wood.
Kilmascally (coill na scaile)—Wood of the shadows.
Kinrush (ceann ruis)—Elevated cape or headland.
Kinturk (ceann sturrice)—Headland of the pinnacle.
Killycolpy (coille colpa)—Full grown heifer's wood.
Killymenagh (coille meadhanach)—Middle wood or wood of the level plain.
Killywoolaghan (coill uailleachain)—Weeping or wailing wood.
Lurgyroe (leargan ruadh)—Red hill side.

Mullaghwotragh (mulach uachtrach)—Upper summit.
Mullanahoe (mulach na h-uamha)—Smmit of the cave.
Mullaghglass (mulach glas)—The green summit.
Sessia (seiseadhach)—A sixth part or divisicn.
Tamlaghtmore (taimhleaght mor) — Large plague monument.
Tamnavalley (tamhnach bhealaigh)—Field of the pass or road.
Trickvallen (tric bhallen)—Uncertain, but it seems to mean a small cup or churn-shaped hill.

PARISH—ARDTREA.

Ardtrea (ard Treagha)—Height of St. Treagha, who founded a church in fifth century—daughter of Partland.
Ballynahone (baile na h-abhann)—Town where rivers rise.
Claggan (cloigean)—Little skull-shaped rocky hill.
Derrygonigan (doire Gonigan)—Gonigan's wood.
Duffless (dubh-lis)—Black fort.
Edernagh (eidir na achadh)—Division of land between townlands; or a central place or field.
Enniskillen (Inis Cethlenn)—Cethlenn's island. She was the wife of Balor a Fomorian chieftain.
Lisboy (lios-buidhe)—The yellow fort.
Liscausy (lios ceassaigh)—The fort of the wickerwork bridge.
Lisnahall (lios na h-olna)—Fort of the wool.
Lurganboy (lurgan buidhe)—Long yellow hill.
Knockanroe (cnoc an ruadh)—Small red hill.
Tullyhurkin (tul an orcain)—Hill of the pigs.
Tullyraw (tula ratha)—Hill of the beech.
Tullyveagh (tul an bheithe)—Hill of the beech.
Tullyweary (tul an mheire)—Hill of mirth where the young assembled for dances and festivities.
Tullyconnell (tul Ui-Connaill)—Connell's hill.
Tievenagh (taobh an achaidh)—Field side.

PARISH—BALLINDERRY.

Ballinderry (baile an doire)—Townland of the oak wood.
Derryerin (doire chaorthainn)—Grove of the mountain ash.

Eglish (eaglaise)—Church land.
Lanaglug (lann na gclog)—The church of the bells.
Mullan (mullan)—Little summit of a low gently sloping hill

PARISH—BALLYCLOG.

Ballyblagh (baile blath)—Town of the flowers or blossoms
Ballyclog (baile an chloig)—Town containing some ancient bell.
Ballyveeny (baile bhinne)—Town of the beann or pinnacle.
Ballywholan (baile faillean)—Town of the little cliff.
Ballynagowan (baile na n-gabhann)—Town of the blacksmiths.
Brigh—A Scotch word Brae, meaning a gentle hill.
Cratley (cruit shliabh)—A hump-backed mountain (Joyce), but it is more likely (creatalach) a place where willow or sally trees grow.
Curglassan (cor glasan)—Round hill of the streamlets.
Drumbanaway (druim beann bhuidhe)—Ridge of the yellow peak.
Drumbulgan (druim Bolgean)—Bolcan's ridge.
Drumkern (druim cairn)—Ridge of the carn or stone heap.
Eary (airghe)—Cultivated or western division of land.
Gortnaskea (gort na sgeach)—The field of the brambles
Kilcoony (coill connaidh)—A wood where firewood is obtained.
Kilsally (coill sailleach)—Wood of the sallow or willow trees.
Killoon (cill shuain)—Wood of repose, where animals rest.
Leck (leac)—Large flat stone or rocky place.
Legmurn (log muirithin)—Swampy pool; or (leg Muirrinne) Murrin's hollow.
Linnyglass (leana glas)—Green swampy meadow.
Oghill (eochaill)—A yew wood.

PARISH—DERRYLORAN.

Aughlish (each laisc)—A place of stables.
Ardcumber (ard chomair)—Hill of the confluence under which rivers meet.

Ardvarnish (ard bhearnais)—High gap in a hill.
Ballymenagh (baile meadhanach)—Middle town.
Ballygroogan (baile Gruagain)—Grogan's wood; or town of the small clamp of turf.
Ballyreagh (baile riach)—Grayish town.
Ballysudden (baile suidin)—Town of the suidin. (Suidin is a plate of oat meal and milk mixed. Much used in the past.
Claggan (cloigean)—Round rocky hill.
Clare (clar)—A level portion of land.
Coolnafranky (cul na bh Franncach)—Back slope of the Frenchmen; or perhaps where rats are numerous.
Coolnahavil (cul na abhaille)—Back of the orchard.
Cranfield (creamhcoill)—Wood of wild garlic
Coolreaghs (cul riach)—Back of the gray ridge.
Cluntydoon (cluainte duin)—The meadow forts.
Craigs (creaga)—Rocky land.
Coolkeeghan (cul Caochain)—Keighen's corner.
Cookstown (cor-a-criche)—Round hill of the boundary.
Drumcraw (druim cruadh)—Hard ridge or ridge of the cattle folds.
Drumgarrell (druim garbh-ghaile)—Ridge of the rough wind.
Drummond (dromann)—Long ridge or hill.
Drumard (drium ard)—High ridge.
Drumearn (druim fhearn)—Ridge of elders.
Derryloran (doire ua Lorain)—Loran's oak wood.
Doorless (dur lios)—Strong fort.
Feegarn (fiodh Gearain)—Garron's wood (O'D.).
Gortalowry (gort Ui Labhradha)—O'Lavery's hill.
Gortin (gort-in)—A little field.
Gortreagh (gort riabhach)—Gray field.
Gallanagh (geal eanach)—White marsh.
Killymoon (coille mughaine)—Mughaine's wood; or (coill mugna) wood of the fat pig.
Kilcronagh (coill cronach)—Wailing or whimpering wood.
Knockaconny (cnoc an chonnaidh)—Hill from which firewood is obtained.
Killycurragh (coille currach)—Marshy wood.
Loy (laigh)—A hill.

Loughry (luachrai)—A place abounding in rushes, cold, stiff retentive soil.
Maloon (magh l-uain)—Plain of the lambs.
Monrush (moin ruis)—Little bog wood.
Moveagh (magh bheitheach)—The plain where beech trees abound.
Sullenboy (sailean buidhe)—Yellow place of sallows or willows.
Tullagh (tulach)—A hill, hilltop or mound.
Tullygare (tulach gearr)—Short hill.
Toberlane (tobar leathan)—Broad large well.
Tullycall (tulach cail)—Hill of the herbs or of the marshy meadows.
Tullywiggan (tulach an bhogain)—Hill of the quagmire or (tulach bhuinn)—Hill of the bulrushes.

PARISH—LISSAN.

Ballinagilly (baile na g-coille)—Town of the woods.
Broughderg (bruagh dearg)—Red overhanging land.
Cluntyganny (cluainte Ganaidh)—Ganny's meadows; or sandy, unproductive meadows.
Coolreaghs (cul riaths)—Back or gray ridges.
Creevagh (craobh achadh)—Field of the wide spreading tree or branchy or wooded land.
Creeve (craobh)—Bushy or wooded land or wide spreading tree.
Davagh (dabhach)—A deep vat-like hollow or cauldron. Could also mean " the field of the oxen," as it has always been noted as a place for fattening bullocks.
Drumgrass (druim greasa)—Ridge of the skirmish.
Dunmore (dun mor)—Big fort.
Feegarren (fiodh Gearain)—Garron's wood.
Lissan (liosan)—A little fort.
Slaghtfreedan (S-leacht Friden)—Freedan's monument. The "s" of leacht is prefixed by metathesis.
Tamneyhagan (tamnach Ui Hagain)—O'Hagan's field.
Tatnagilta (tate na n-gioltagh)—Land measure (60 acres) of the reeds.
Unagh (uaimhneach)—A lonely place, a bleachgreen.

PARISH—TAMLAGHT.

Aghaveagh (achadh a' bheitha)—Field of the birches.
Coagh (cuas)—A cave or grotto; also a low-lying hollow place.

Drumard (druim ard)—High ridge.
Drumconway (druim con bhuidhe)—Conway's ridge.
Mullaghtironey (mullach tir Eoghain)—Owen's hill top.
Urbal (earbail)—Tail end projection of land.
Tamlaght (tamlacht)—A burial ground.

PARISH—DONAGHENRY.

Annaghone (eanach Eoghain or ahhann)—Owen's marsh or river marsh.
Aghalarg (aghadh leirge)—Hillside of the slopes or battle.
Ardpatrick (ard Padraig)—Patrick's height.
Cahoo (ceath)—Exposed place, not protected from storm.
Cloughfin (cloch fhionn)—White quartz like stony land.
Coolatinny (cul a' tsionnaigh)—Hill corner of the fox.
Donaghenry (domnach fhainre)—Sloping church land.
Donaghey (domnach)—Church land.
Dooragh (dubh rath)—Black rath or fort.
Drumagullion (druim na g-cuillion)—Ridge of the holly bushes.
Drumey (druim)—Ridgy land; or (druim Aodha) Hugh's ridge.
Drumgormal (druim gorm maol)—The blue bare ridge.
Gortagammon (gort na g-caman)—Field of the camans or hurleys.
Gowshill (gabha)—The smith's hill.
Galvally (gall bhaile)—The foreigner's (English) town
Gortacloghan (gort a' chlochain)—Field of the stepping stones along a river.
Gortatray (gort a t-srae)—Field of the mill race.
Innovall (in bhear mhall)—A slow flowing estuary.
Killymurphy (coille murchadha)—Murphy's wood.
Lislea (lios liath)—The gray fort.
Letterclery (leiter Cleirigh)—Clery's or the clerics wet or spewy hillside.
Liskittle (lios Citail)—Kittal's fort (O'D.).
Lisneight (lios Neacht)—Neaght's fort (O'D.).
Lurgy (luragh)—A long hill.
Mullaghmoyle (mullach maol)—Bald or bare hilltop.
Mullantain (mullan t-sian)—Hill of the fairy finger or foxglove.
Mullaghmoyle (mullach maol)—Bare hilltop.
Ross (ros)—A promontory or land jutting out.

Rouskey (rusgaidh)—Marshy rough or fenny land.
Rouskeyroe (rusgaidh ruadh)—Red marshy or fenny land.
Shankey (sean ceadh)—An old quagmire.
Sherrigrim (searach dhruim or sceiri dhruim)—Ridge of the foals or stony ridge.
Soarn (sorn)—A place of a lime kiln.
Templereagh (teampoll riadh)—Gray church.
Tullagh Beg (tulach beg)—Little hill.
Tamneylennon (tamnach Lennon)—Lennon's green field.
Tullagh Mor (tulach mor)—Great or big hill.
Tullyfaughan (tulach fothain)—Sloping hill top or hill of the strife.
Tullylig (tulach luig)—Hill of the hollow.
Unicks (unach)—A washing or bathing place.
Urbalreagh (urbal riadh)—Grey projecting land resembling a tail.

PARISH—DESERTCREAT.

Annaghquin (eanach cuinn) Quinn's marsh.
Allen (aillin)—A little cliff or beautiful spot.
Annaghavil (eanach abhaille)—Marsh of the orchard.
Annaghananam (eanach an anama)—Marsh of the soul or ghost.
Annaghmore (eanach mor)—Great marsh.
Annaghteige (eanach Teigue)—Teigue or Timothy's marsh.
Ballynacroy (baile na cruaiche)—Town of the round hill.
Ballynakilly (baile na coille)—Town of the wood.
Ballymully (baile mullaigh)—Town built on summit of a hill.
Bardahessiagh (barr an da coille)—Summit of the two sessiaghs, or sixth part of a division of land.
Cady (ceide)—A hill or hilly land.
Carnenny (carn eanaigh)—The rocky marsh.
Derryraghan (doire rathain)—Ferny woodland, or of the signal fire or beacon.
Drumraw (druim ratha)—Ridge of the fort.
Desertcreat (diseart crioth)—Waste place of the territories.
Donaghrisk (domnach riasca)—Church of the moor.

COOKSTOWN UNION.

Derrygortanea (doire gort an fhiaidh)—Grove of the deerfield.
Drumballyhugh (druim baile Aodh)—Townland of Hugh's ridge.
Derryash (doire thais)—Soft oak grove.
Edendoit (eadan docht)—Hillbrow of the unprofitable land.
Finvey (fionn mhaigh)—Fair or beautiful plain.
Gortagowan (gort a' ghabhann)—Field of the blacksmith.
Gortavilly (gort a bhile)—Field of the ancient tree.
Gortfad (gort fada)—Long field.
Gorticar (gort a' charra)—Field of the stone or causeway.
Grange (grainseach)—A grange or the outoffices attached to a farm house.
Gortindarragh (gort in darach)—Little field of the oak.
Galcussagh (geal coise)—Low lying land covered with white flowers.
Gortivale (gort aidhbheil)—A very large field or opening into a wood or valley.
Gortigar (gort an chairr)—Rocky field or rocky surface; or of the causeway.
Killygarvan (coille a garbhain)—Wood of the rough land.
Killyneedan (coill an eadain)—Wood of the brow or low hill.
Kiltyclogher (coillte clochair)—Woods of rocky ground.
Killycolp (coill colpa)—Wood of the fully grown heifers
Kiltyclay (coillte cleithe)—A wood from which poles or hurdles are obtained.
Knockavaddy (cnoc a mhadaigh)—The dog's hill.
Lammy (leamhaidh)—Land of elms.
Legacurry (lag-a-churraigh)—Hollow of the cauldron or pit.
Lisnanane (lios na n-ean)—The fort of the birds.
Mullaghshantullagh (mullach sean tullach)—Summit of the old hill.
Moneygaragh (muine garbh)—Rough shrubbery.
Moree (magh ruadh)—Reddish coloured plain.
Moynagh (magh n-each)—Plain where horses graze.
Mollynure (mul-an iubhair)—Hilltop or summit of the yew
Moymore (magh mor)—The great plain.

Oughterard (uachter ard)—The upper height.
Shivey (sithbhe)—Fort or a fairy palace.
Skenarget (sciath an airgid)—Land of the silver thorn.
Skenahergney (sceath na airgne)—Bush of plunder.
Sessiagh (seisadhach)—Sixth part or division.
Tullyreavy (tulagh riabhach)—Grey field or hill.
Tully (tulach)—A gentle hill.
Tullyard (tulach ard)—High hill.
Tullylaghan (tulagh leathan)—Wide roomy hill.
Tirnaskea (tir na sceagh)—District of the thorns.
Tolvin (tulach bhinn)—Beautiful hill.
Tullydonnell (tulach Domhnaill)—O'Donnell's hill.
Tullaghogue or Tullyhog (tulaigh og)—A little hill. According to Joyce, Tullaghogue signifies "the hill of the youths," where games were carried on and also where the inauguration of the O'Neill chieftains took place. Until fifty or sixty years ago there was a yearly gathering of the youth of the district commemorating the games of ancient Ireland.

PARISH—KILDRESS.

Ballinasollus (baile na solus)—Town of the ford of the torches.
Beleevna (beal aoibhneach)—Beautiful river mouth; or (beal buibheanna) the river mouth of the herbs.
Beltonanean (bailte na n-ean)—Town of the birds; or (Beltaney) a place where Druidic fires were lit on May Day.
Beaghbeg (beitheach beg)—Little birch wood.
Beaghmore (beitheach mor)—Large birch wood.
Clare (clar)—A plain or level tract of land.
Clontyferagh (cluainte fearach)—Meadows producing an abundant rich grass.
Corchoney (corr connaidh)—Rabbit hill of the brushwood.
Corvanaghan (cor Bhanachain)—Bannigan's round hill.
Cavanoneill (cabhan Ui Neill)—O'Neill's hillside; cabhan also means a hollow.
Cloughfin (cloch fionn)—Land covered with white stones quartz.
Derrinleagh (doire liath)—Small gray oak wood.
Drum (druim)—A ridge.
Drumnacross (druim na croise)—Ridge of the cross.

Doons (duin)—Earthen forts.
Drumnagloch (druim na gloch)—The hill of the stones.
Drumnamalta (druim na mealta)—Ridge of the lumps or hillocks.
Dunamore (dun mor)—Big fort.'
Dungate (dun gead)—Fort of the white or cosy spot of land.
Drumshambo (druim sean bhoithe)—Ridge of the old hut or tent.
Evishbrack (ebhis breac)—Speckled mountain pasture.
Evishacrancussey (eibhis na crann coise)—Wild growth of the tree stumps.
Evishanoran (aibheis a fhuarain)—Pasture surrounding a spring well.
Glenarney (gleann airneadh)—Glen of the sloes.
Gortreagh (gort riabhach)—Gray field.
Gortin (goirtin)—A small field.
Gortnagross (gort na gcros)—Field of the cross.
Gortaclady (gort an chladaigh)—Miry or muddy field.
Glasmullagh (glas mulach)—Green summit.
Keenaghan (caeneaghan)—Small mossy field.
Kildress (coill dreasog)—Wood of the briars or brambles (O'D), but more correctly, I think (coill-dar-eas(a)) oak wood of the waterfall.
Knockaleery (cnoc a Laoirghre)—O'Leary's hill. O'Leary was the chieftain of the district and was buried in Tattykeel.
Kinnegillion (ceann na g-cuillean)—Head of the holly bushes.
Killeenan (coill lionbhan)—Church of the hard bare patch of land.
Killucan (coill lucain or lochan)—The wood of the marsh mallow, or it might mean the wood of the little lakes.
Legnacash (leg na ceasaighe)—Hollow of the kish or wicker bridge or boggy road.
Magheraglass (machaire glas)—The green plain.
Mackenny (magh caoine)—The smooth level and beautiful plain.
Mintober (mion tobar)—The small or tiny well.
Meenanea (min an iodha)—Level place of the yew tree or ivy.
Meenascallagh (min na scaile)—Field of the shades or shadows.

Moboy (magh buidhe)—Yellow plain.

Murnells (murtha Neill)—O'Neill's mounds.

Orritor (oirthear)—Land having an eastern aspect. Some say it means a district lying along a river.

Strews (sraoth)—River meadows or streamlets or a mill race.

Tamnaskeeny (tamnach scine)—Knife-shaped field or pointed hillock.

Tattykeel (tate caol)—Narrow tate or division of land.

Terrywinny (tir a bhainne)—The milk producing district.

Tamlaght (tamhlaght)—A plague monument or burial mound.

Tirmacshane (tir MacShean)—M'Shane's district or country.

Tullyroan (tulach roan)—Red hill summit.

Tulnacross (tulach na croise)—Hill of the cross.

Teebane (taobh ban)—Sunny or bright side. Joyce gives it (tiogh ban)—White house.

DUNGANNON UNION.

PARISH—POMEROY.

Aghafad (achadh fada)—The long field.
Aghnaskea (achadh na sgeach)—The field of the whitethorns.
Altmore (alt mor)—Big glen.
Ballymacall (baile MacCaul)—M'Caul's town.
Claggan (cloigean)—Round rocky hill.
Coolmaghry (cul machaire)—Plain in the corner (of a wood, etc.).
Creeve (craobh)—A wide spreading tree.
Corkhill (cor choill)—Round hill of the hazels.
Crossdermot (crois Dhiarmuda)—Dermot's cross.
Curlonan (cor loanain)—Hill of the boasting.
Cavanacaw (cabhan na caithe)—Hill of the chaff, where corn was cleaned without fans.
Corrycroar (corr an chreabhair)—Hill of the woodcock.
Camaghy (cam achadh)—The crooked field.
Cappagh (ceapach)—Tilled or cultivated land; or land abounding in tree stumps (O'D).
Cornamaddy (Corr na madadh)—Round hill upon which dogs congregate.
Crannogue (cranog)—A place abounding in trees or a house made of wickerwork erected on stakes in a little lough or lake.
Cavanakeerin (cabhan an chaorthainn)—Round hill of the mountain ash
Crosscavanagh (cros Cavanagh)—Cavanagh's cross (O'D.).
Drummond (dromann)—A long hill.
Dungororan (dun Mhic Furahrain)—Fort of the son of Foraran.
Dernanaught (doireanach)—Woody face of a hill.
Drumconor (druim Chonchubhair)—Connor's ridge.
Gortavoy (gort a bheithe)—Field of the birch trees.
Gortindarragh (gortin darach)—Little field of the oak.
Galbally (gall baile)—Town of the foreigners or strangers.
Glenbeg (gleann beg)—A small glen.
Glenburrisk (gleann burrais)—Glen of the caterpillars.
Gortnagola (gort na ngabhla)—Field of the forks where rivers or streams meet.

Gortnagarn (gort na g-carn)—Field of the stone heaps. parsons.

Kerrib (eirb)—A section or portion of a division of land

Kilmacardle (coill MacArdle)—M'Ardle's wood.

Kilmore (coill mhor)—The big wood.

Knocknaclogha (cnoc na gclocha)—The stony hill.

Killey (coill Aodha)—Hugh's wood, or a woody place.

Lisnagleer (lios na g-cliar)—Fort of the clergy.

Lurgylea (lurga liath)—Long gray hill.

Lurganeden (lurgan eadan)—Long hill face.

Mulnagore (mulach na ngabhar)—Summit of the goats.

Munderrydoe (moin an doire dhuibh)—Bog of the black oak wood.

Skea (sgeach)—A lone thorn bush.

Sessiadonaghy (seiseadhach Donnachadha)—Donaghy's sixth part or division

Shanmaghry (sean-mhachaire)—The old plain.

Tulnagall (tulagh na n-gall)—Hill of the foreigners.

Tanderagee (toin le gaoith)—A place whose "backside" faces the wind.

Turnabarson (tor na bparson)—Round tower of the parsons.

PARISH—CLONFEACLE.

Annagh (eanach)—Marshy ground.

Anagasna (achadh na g-casna)—Field of the wood chips

Altnavannog (alt na bhfeannog)—Cliff or hill of the scaldcrows.

Ballymackleduff (baile Mac Giolla dubh)—M'Elduff's town.

Benburb (beann borb)—Proud or bold cliff.

Boland (bo lann)—An enclosure for cows.

Brossley (brus lighe)—Grove of the dust or broken pieces.

Broughadowey (bruach a dubhaigh)—Brink of the black bog.

Carrowbeg (ceathramhadh beag)—Small quarter.

Carrowcolman (ceathramhadh Colman)—Colman's quarter.

Coolhill (cul-choill)—Back wood.

Crew (craobh)—Wide spreading tree.

Crubinagh (cruibeneach)—Land formed like claws or talons.

Curran (cuirrin)—A small pit.
Cadian (ceidin)—Small hill, level at top.
Carrycastle (carraig a chaisil)—Rock of the castle; stone fort or church rock or boundary wall.
Clogharny (clocharnaidh)—A stony place.
Crossteely (cros an tighe liaith)—Cross of the grey building.
Clonbeg (cluain beag)—Small meadow.
Clonmore (cluain mor)—Big meadow.
Clonteevy (cluain taoibhe)—Lawn of the hill side.
Culkeerin (cuil caorthainn)—Corner where rowan trees grow; or the backside bog.
Culrevog (cuil rubha)—Hill back producing the herb rue.
Coolcush (cul coise)—Hill back of the foot shaped land.
Cormullagh (cor mullach)—Round summit.
Derrycreevy (doire craoibhe)—Oak wood of the branching tree.
Derryfubble (doire phobail)—Wood of the congregation where religious ceremonies were held.
Derrygoonan (doire Ua gCuanain)—O'Coonan's wood.
Drumay (druim Aodha)—Hugh's ridge.
Drumfluch (druim fluich)—The wet ridge.
Drumgose (druim g-cuas)—Ridge of the cave.
Drumnastrade (druim na sraide)—Ridge of the highway.
Drumseark (druim searc)—Hill where lovers meet.
Derrylatinee (doire na teine)—Wood of the fires.
Drain (drainn)—Great round hill.
Drumnamoless (druim na mol-leasa)—Ridge of the solid fort.
Drumskinny (druim sceine)—Knife shaped ridge.
Derryoghill (doire eochaille)—Wood of the yew trees.
Drumanney (druim manaigh)—Monk's ridge.
Drumderg (druim dearg)—Red ridge.
Drumgart (druim geart)—Ridge of the fertile tillage plots.

Drumgold (druim guail)—The coal ridge.

Drumgrannon (druim greanach)—Small gravelly ridge; or (druim gronndain) ridge of the lamentation.
Drumlee (druim laoigh)—Ridge of the calves; or (druim liath) grey ridge.
Drummond (dromann)—A long hill.

Derrycreevy (doire craoibhe)—Wood of the branching trees.
Donnydeade (dun a deid)—Fort of the tooth shaped hill or fine embankment.
Derrygortrevy (doire guirt riabhaigh)—Wood of the grey field.
Drumgormal (druim gormmaol)—Blue bare ridge or (gabhair maoil) of the hornless goat.
Drumnasaloge (druim na saoileog)—Ridge of the sallows or willows.
Dunamony (dun an mhuine)—Fort of the shrubbery.
Finelly (fionn aille)—Fair cliff.
Garvaghy (garbh achadh)—Rough field or pasture.
Gort (gort)—A field.
Gortmerron (gort Merron)—Merron's field.
Gorestown (gabhar)—Goat's town, where goats are plentiful.
Grange (grainseach)—A monastic granary or place where grain is stored.
Kilnagrew (coill na gcraobh)—Wood of the branching trees.
Knockarogan (cnoc a ruagain)—Hill of defeat. A battle must have been fought here; or hill of the sharp dry breeze.
Knocknacloy (cnoc na cloiche)—Hill of the remarkable stone.
Killybracken (coille Breacain)—Bracken's wood.
Kilnacart (coille na ceardchan)—Wood of the forge.
Lisbancarney (lios ban Chearnaigh)—Carney's white fort.
Lisbanlemneigh (lios ban leim an eich)—White fort of the horse leap.
Lisduff (lios dubh)—Black fort.
Lisgobban (lios gobbain)—Little snout like fort.
Lisnacroy (lios na croiche)—Fort of the gallows, where hangings were carried out.
Lisroan (lios Ruadhan)—Rowan's fort or red fort.
Listamlet (lios tamhlachta)—Fort of the burial place.
Lisdermot (lios Diarmada)—Dermot's fort.
Lismulrevy (lios Maoil Riabhaigh)—Mulreavy's fort; or fort of the gray top or hillside.
Legilly (lag gile)—Bright pool or hollow.
Lissan (liosan)—A little fort.

Mossmore (moss mor)—The great moss; or (mas mor) great fat hill.

Moyard (magh ard)—A high plain.

Mullaghdaly (mullach dala)—Hill where meetings were held.

Mullycarnan (mullach carnain)—Summit of the stone heap.

Mullaghlongfield (mullach leamh choille)—Hill of the elm wood.

Mullycar (mullach carr)—Summit of the rocks.

Moy (magh)—A plain or level extent of land.

Mullaghboy (mullach buidhe)—The yellow summit.

Mullaghmossog (mullach measog)—Hill of the berries.

Moycashel (magh caisil)—Plain of the stone fort.

Mullybrannon (mullach branair)—Summit of the fallow land or chief's ridge.

Mulboy (mullach buidhe)—The yellow summit.

Roan (ruadhchan)—Reddish coloured land.

Sanaghanroe (seanachan ruadh)—Red fox land; or (scan ath-an ruaidh) old fort of the red men or rue

Sassiamagarroll (seiseadhach Mhic Cairill)—M'Carroll's sixth part or division.

Stilloga (stealloga)—Little strips of land.

Syerla (suidhe iarla)—House or seat of the earl.

Shanmoy (sean mhagh)—Old plain.

Seyloran (soil Odhrain)—Oran's seed or race (O.D.). Joyce has it Suidhe Uarain—Seat of the cold spring

Strangmore (strang mor)—Great ford. Strang is a Danish word signifying ford.

Terryscollop (doire scolb)—Oak wood of scollops or rods used for thatching.

Tullygowney (tulaigh connaidh)—Hill of the firewood.

Terryglassog (doire glasog)—Oakwood of the small green plain; or (glasoige) the "wagtail's" grove.

Tullygiven (tulaigh Geimhin)—Given's hill.

Tobermessan (tobar Masann)—Mason's well, or (muisean) well of the primroses.

Tullydowey (tullach Dubhtaigh)—Duffy's hill.

Tullylearn (tullagh liath fearainn)—Hill of grey land.

Turleenan (tur Lionain)—Leenan's bush or tower; or (an lionan) hill of the short flax.

Tyhan (toigh Sheain)—John's house.

PARISH—AGHALOO.

Aghaloo (achadh Lugha)—Lewy's field.
Annaghmore (eanach mor)—Great marsh.
Aghenis (each inis)—Horse's holm or island.
Anacramp (eanach creamha)—Marsh of the wild garlic
Annaghroe (eanach ruadh)—The red marsh
Ards (aird)—Heights or small hills.
Annaghsallagh (eanach salach)—Dirty or miry marsh.
Annagh (eanach)—A marshy place.
Bohard (both ard)—High hut.
Ballagh (bealach)—A pass or leading road.
Ballyboy (baile buidhe)—Yellow town where the soil is of that colour.
Ballyvaddy (baile an mhadaidh)—Dog's town.
Carricklongfield (carraig leamhcoille)—Rock of the elm wood.
Crilly (crithleach)—A shaking bog.
Croughill (creamh choill)—Wood of the wild garlic.
Caledon—The old Irish name is Ceannard, high head.
Cavanbuoy (cabhan buidhe)—Hard, dry, yellow hill.
Culligan (cuilleagan)—A hazel shrubbery.
Curlagh (cor leachta)—Round hill of the monument.
Creevelough (craobh locha)—Spreading tree of the lake.
Cumber (cumar)—A confluence or meeting of rivers.
Derrycourtney (doire Cuairtain)—Courtney's wood.
Derrygooly (doire guala)—Wood of the charcoal.
Derrykintone (doire cuin tamhain)—Wood of the hill of the tree trunks.
Dromore (druim mor)—The big summit.
Drumess (druim easa)—Ridge of the waterfall.
Dunmacmay (dun Mhic Mheith)—M'Veigh's fort.
Dyan (daingean)—A stronghold.
Derrylappin (doire Ui Lapain)—O'Lappin's wood.
Drumearn (druim fhearn)—Ridge of the elders.
Drummond (dromann)—A long ridge or hill.
Edengeerach (eadan na g-caorach)—Hill side of the sheep.
Enagh (eanagh)—A marshy, low-lying place.
Finglush (fionn ghlaise)—a bright streamlet.
Glasdrummond (gleas dromann)—Long green hillside.
Glendavagh (gleann dabhach)—Glen of the deep wells or vats.

Glenkeen (gleann caoin)—Beautiful glen or valley.
Glenarb (gleann earb)—Glen of the roebucks.
Guinness (gineadha)—Fertile spots of land.
Glencrew (gleann craobh)—Glen of the woody or branchy trees.
Knocknaroy (cnoc an roigh)—King's hill, or perhaps hill of the executions.
Kedew (ceide)—A green hill or level hillock.
Kilgowney (coill gamhna)—Wood of the calves or strippers.
Kilsampson (coill Samsoin)—Sampson's wood.
Knockaginny (cnoc a gineadha)—Hill of fertile land
Killynaul (coill an fhail)—Wood of the hedge or enclosure.
Kilmore (coill mhor)—Big wood.
Kilsannagh (coill seanach)—Wood inhabited by foxes.
Legane (lagan)—A small hollow or dell.
Lairakeen (laithreach caoin)—Beautiful situation or site.
Lismulladown (lios Maoil Duin)—Muldoon's fort.
Mullaghmore (mullach mor)—Big summit.
Mullaghmossagh (mullach maiseach)—Beautiful summit.
Mullynaveigh (mullaigh na bfiadh)—Hill of the ravens.
Mullintor (mul an tuair)—Summit of the bleach green; or perhaps the residence of a local chieftain.
Mullycarnan (mullaigh carnan)—Summit of the stone heaps.
Mulnahorn (mul na h-eorna)—Summit of the barley; or (mul na charnain) of the stone heaps.
Mullyneill (mullaigh Neill)—Neill's summit.
Rehaghy (raith Eachadha)—Eochy's fort
Ramaket (rath Mhic Ceit)—Fort of Keith's son.
Stragrane (srath granda)—Ugly rath or holm.
Tullyblety (tullaigh bleite)—Hill of the grinding where oats were ground.
Tannaghlane (tamhnach leathan)—Wide green field.
Tullynashane (tullaigh na sian)—Stormy hill top.
Tullyremon (tullaigh Reamoin)—Redmond's hill.
Tannagh (tamhnach)—A green fertile field.

PARISH—DONAGHMORE.

Aghintober (achadh an tobair)—Field of the small spring well.

Aghnagar (achadh na g-corr)—Field of the cranes or herons.
Altaghlushin (alt an ghlasain)—Glen of the green herbs
Annaghmackeown (eanach MacCozhain)—M'Keown's marsh.
Aughlish (each laisc)—A stable for horses.
Agharan (achadh raitheain)—Ferny field.
Aghareany (achadh raithnighe)—Ferny field.
Annagh Beg (eanach beag)—Little marsh.
Annaginny (eanach gaineadh)—Fertile marsh.
Ballyward (baile an Bhaird)—Ward's town.
Ballysaggart (baile sagairt)—Priest's town.
Ballybray (baile breagh)—Good looking town.
Clonavaddy (cluain a mhadaidh)—Dog's meadow.
Cullinfad (cullion fada)—long strips of land covered with holly.
Cullenramer (cuillion ramhar)—Thick or close holly wood.
Clananeese (cluan na nise)—A meadow of potters clay.
Creevagh (craobhach)—Bushy or wooded land.
Dernaseer (doire na saor)—Grove of the carpenters or mechanics.
Dristernan (driosarnan)—Blackthorn shrubbery.
Dredolt (droichead alt)—Bridge of the steep glen side.
Derryalskea (doire aile sgeach)—Oak grove of the briar cliffs.
Derryhoar (doire ur)—New oak grove, not long planted
Drumhirk (druim thuirc)—Ridge of the wild boar.
Drumnafern (druim na bhfearn)—Ridge of the alder trees.
Drumreany (druim raithnighe)—Ferny ridge.
Derryveen (doire mhin)—Smooth or leel oak wood.
Donaghmore (domhnach mor)—The great church.
Drumbearn (druim bearnain)—Ridge of the gap or chasm.
Derrykeel (doire caol)—Narrow wood.
Edenacrannon (eadan crannain)—Hill brow of the trees
Eskragh (eiscreach)—Abounding in eskers, or long, low, gravelly hills.
Finulagh (fionn loch)—White lough.
Feroy (feur fhaithche)—A grassy lawn.
Foygh (faithche)—A fair green or lawn.
Gorey (guaire)—Abounding in long grass.

Gortlenaghan (gort Luineachan)—Lenaghan's field.
Glasmullagh (glas mullach)—Green summit.
Glenadush (gleann a duis or gleann da dhos)—Glen of the bushes or glen of the two bushes.
Garvagh (garbh achadh)—Rough field.
Gortnaglush (gort na glaise or gort na h-eaglaise)— Field of the streamlet or field of the church.
Kilnaslee (coill na slighe)—Wood of the road or pass.
Killylevin (coill Ui Leibhin)—Levin's wood.
Killyliss (coill a leasa)—Wood of the fort; or (coill a' l-easa) of the waterfall.
Killymaddy (coill a mhadaidh)—Dog's wood.
Killymoyle (coill na maoile)—Wood of the hornless cow
Killyquinn (coill Ui Chuinn)—Quinn's wood.
Killycavanagh (coill cabhanaighe)—Wood of the small hills.
Killyharry (coill an charraigh)—Rocky wood.
Lisgallon (lios gallain)—Gallon's fort or perhaps the fort of the standing stones.
Lisnamonaghan (lios Monchain)—Monaghan's fort.
Lisnahull (lios na h-olna)—Fort of the wool.
Lisboy (lios buidhe)—Yellow fort.
Lisnagowan (lios na n-gabhann)—Fort of the blacksmiths.
Moghan (muchan)—Flooded bogland.
Mullaghbane (mullach ban)—White summit.
Mullaghadrolly (mullach drolaigh)—Summit of the hooks (O'D.). Joyce has it—Summit of the windings.
Mullaghanagh (mullach anatha)—Summit of the fort.
Mullaghacreevy (mullach craobhaighe)—Summit of the wide-spreading tree.
Mullaghconor (mullach Conchubhar)—Connor's summit.
Mullaghmore (mullach mor)—The big summit.
Mullaghrodden (mullach roidin)—Summit of the little road.
Mullycrunnet (mulaigh cruithneachta)—Summit of the wheat.
Mullygruen (mulaigh groighen)—Hill where horses or brood mares graze.
Reclaim (rath claon)—Inclining or leaning fort.
Reaskcor (riasc corr)—Marsh where cranes congregate

Reloagh (raileach)—Abounding in Oaks (O'D.)
(reidheagh)—A flat or level plain (Joyce).

Reaskmore (riasc mor)—Big morass or swampy place.

Stakernagh (stacan achadh)—A place where stakes or posts were cut.

Tullyaran (tulaigh bhfearn)—Hill of the alder trees; or (tulaigh ghearan)—Hill of the horses (O.D.).

Terrenew (tir an fheadha)—District of the woods.

Toomog (tuamog)—A burial mound.

Tullyallen (tulaigh alainn)—Beautiful hill.

Tullydraw (tulaigh idir dha rath)—Hill or field between two forts.

Tullyleek (tulaigh lig)—Stony hill.

Tullynure (tulaigh n-iubhar)—Hill of the yew trees.

PARISH—TULLYNISKIN.

Aghakinsellagh (achadh cinn saileach)—Field of the hill of willows (O'D.). Kinsella's field (Joyce).

Ballymenagh (baile meadhanach)—Middle town.

Bloomhill and Bloomry—Dr. O'Donovan seemed to think these are English names.

Creenagh (crionach)—Withered land where trees, etc., are decaying.

Cullin (cullion)—Holly bearing land.

Curran (carn)—a heap of stones or a stony place; or (cuirin)—a deep pit or depression.

Derry (doire)—An oak grove.

Derrywinny (doire mhuine)—Grove or wood of the shrubbery.

Doras (dubh ros)—Black peak or wood.

Drumard (druim ard)—High ridge.

Drumey (druim Aodha)—Hugh's ridge.

Drumreagh Etra (druim riadh iochtarach)—Lower grey ridge.

Drumreagh Otra (druim riadhuachtarach)—Upper grey ridge.

Edendork (eadan na d-torc)—Hill brow of the hogs.

Farlough (fuar loch)—Cold lake.

Glencon (gleann con)—Glen of the hounds.

Gortin (gort-in)—A little field.

Gortgonis (gort gabha an easa)—Smith's field, which is situated near a waterfall; or (gort chonaidh)—Field of the firewood.

Gortnaskea (gort na sgeach)—Field of the briars or thorns.
Mineveigh (min na bhfiach)—Field of the ravens.
Mullaghmarget (mullach margaidh)—Hill on which a market was held.
Quintinmanus (cointin Maghnus)—Manus's controversy or disputed land or territory.
Sessia (seiseadhach)—Sixth part or division.
Stughan (stuacan)—A little rocky point or projection.
Tullyniskin (tulaigh eascon)—Hill of the water-dog; (tullaigh eascaine) of the curse.

PARISH—DRUMGLASS

Coolhill (cul choill)—Back wood (most southern town-land in the parish).
Cornamucklagh (cor na muclach)—Round hill of the piggeries.
Congo (cumhanga)—A narrow strip of land or water.
Creevagh (craobhach)—Bushy or wooded land.
Drumhariff (druim tairbh)—Bull's ridge.
Dervaghadoan (doire achadh duinn)—Grove of the brown field.
Drumglass (druim glas)—Green ridge.
Drumcoo (druim cuaiche)—Cuckoo's ridge.
Dungannon (dun Geannain)—Gannon's fort; Gannon was the son of Caffa, a druid who lived in the first century.
Gortmerron (gort Mearain)—Merron's field.
Killylack (coill na leac)—Hill of the flag stones.
Killyneill (coille Neill)—O'Neill's wood.
Kingarve (ceann garbh)—Rough head or hill.
Killymeal (coill na maoile)—Wood of the hornless cow.
Killybrackey (coill breacaigh)—Wood of speckled land.
Lisnaclin (lios na clinge)—Fort of the bell chime.
Lurgaboy (lurga buidhe)—A long yellow hill.
Mullaghadun (mullach an duin)—Summit of the dun or fort.
Ranaghan (raithneachan)—Land abounding in ferns.
Ross Beg (ros beag)—A small promontory or wood.
Ross More (ros mor)—A large promontory or wood.
Tullycullion (tulaigh cuilinn)—Hill of the holly-bush.
Tullygun (tulaigh na g-con)—Hill of the hounds.
Tullyodonnell (tulaigh Domhnaill)—O'Donnell's hill.

PARISH—KILLYMAN

Annaghbeg (eanach beag)—Little marsh.
Ballynakilly (baile na coille)—Town of the wood or church.
Bogbane (bog ban)—Soft green field.
Bovean (both mheadhan)—Central hut.
Bernagh (bearnach)—A gapped hill or place.
Cohannan (coimh ionain)—A level surface.
Corr (cor)—A round hill.
Corrainy (cor raithnigh)—Round ferny hill.
Cavan (cabhan)—a round hill.
Coash (cuas)—A place containing a cave.
Culnagor (cul na g-corr)—Hill back of the cranes or herons.
Culnagrew (cul na g-craobh)—Hill back of the bushes.
Derrygalley (doire gallaigh)—Wood or grove of the rocks.
Derrymeen (doire min)—Wood of the mountain meadow or smooth ridge.
Drumard (druim ard)—High ridge.
Drumaspill (druim aspaill)—Ridge of the apostle.
Drumcrow (druim cruadh)—Hard ridge; or (druim crodh) ridge of the cattlefolds (O'D.).
Drumenagh (druim meadhonach)—Middle ridge.
Drumhorrick (druim na comhairce)—The boar's ridge.
Drumkee (druim ceath)—Ridge having a western aspect, exposed to showers; or (druim ceo) foggy ridge.
Dreemore (druim mor)—Big ridge or hill face.
Drummuck (druim muc)—Ridge of the pigs
Dungorman (dun Gormain)—Gorman's fort.
Gortrea (gort riabhach)—Grey field.
Gortshalgan (gort sealgain)—Field of the sealgan (an edible herb).
Killyman (coill na meathan)—Wood of the saplings, twigs or soft growths.
Keenaghan (caenachan)—A mossy place.
Kinego (ceann an ghabhann)—The hill of the smith.
Laghey (liatha)—Grey lands (O'D.). Slough or miry place (Joyce).
Leiderg (leath dearg)—Red half or place.
Lisnahoy (lios na h-aithche)—Fort of the lime kiln.

Mullaghteige (mullach Teige)—Teague or Timothy's summit.
Moyroe (magh ruadh)—Red plain.
Tamlaghtmore (tamhlachtmor)—Big burial place.
Tartalaghan (tearc lochan)—Dry pool or small lough.
Tempanroe (tiompan ruadh)—Red hillock.
True (an tsruth)—A snout or point of land.

PARISH—KILLEESHIL.

Aghnahoe (achadh na h-uaighe)—Field of the grave or cave.
Aghaginduff (achadh cinn duibh)—Field of the black head.
Ballynahaye (baile na h-aithe)—Town of the lime kiln.
Bockets (buacaide)—A round hill.
Cranlome (crann lom)—Place of leafless or bare bushes
Cullentra (cul an t-sratha)—Hill back of volley along riverside.
Cabragh (cabrach)—Bad, waste land.
Cloontyclevin (cluainte Sleibhin)—Slevin's meadows; or (cluainte clamhan) meadows of the baregrass.
Cloontyfallow (cluainte Fallamhain)--Fallon's meadows or (cluainte fallamh) empty or bare meadows.
Coolhill (cul coille)—Back side of the wood, most southern townland in parish.
Dergenagh (dearganach)—Red marshy ground.
Drumfad (druim fada)—Long ridge.
Ennish (innis)—River meadow, smooth pasture along a river.
Eskragh (eiscreach)—Full of eskers or long low hills.
Farriter (fearan iochtrach)—Lower lying land.
Fasglasagh (fas glaiseach)—Wilderness of streamlets.
Glencull (gleann choill)—Hazel glen.
Killeeshil (coill iseal)—The lower wood.
Lisfearty (liosfearta)—Fort of the graves.
Lurgacullion (lurg a chuilinn)—The long hill of the holly.
Mullysilly (mullach sailigh)—Hill of the willows.
Mullyrodden (mullagh rodain)—Summit of the little road; or (mullach rodan) summit of the ferruginous or spa spring.
Mulnahunch (mullach na h-uinsinne)—Hilltop of the ash trees.

Tullyvannon (tul an bheannain)—Summit of the little peak.

PARISH—CARNTEEL.

Annaghbeg (eanach beag)—The little marsh.
Aughnacloy (achadh na cloiche)—Field of the stone or stony field.
Armalughey (ard maol Eachadha)—Eoghy's bare hill.
Ballynapoteogue (baile na bpotog)—Town of the entrails.
Belrath (baile ratha)—Town of the fort or rath.
Branny (brannaidh)—Pens or folds for sheep.
Cavankilgreen (cabhan na coille greine)—Round hill of the sunny wood.
Cavan O'Neill (cabhan O'Neill)—O'Neill's round hill.
Cravenny (craobh Eanaidh)—Enny or Ethna's spreading tree.
Carnteel (carn t-Siadhail)—Shiel's carn.
Corderry (cor a doire)—Round Hill of the oak wood.
Cranslough (crann loch)—Lough or lake overhung with bushes.
Derrycush (doire coise)—Bottom wood land.
Derrycreevy (doire craoibhe)—Oak wood of the branching trees.
Doolargy (dubh leargaidh)—Black slope or hillside.
Drumaslaghy (druim na sleamhach)—Ridge of the elm trees.
Dernaborey (doire na boirche)—Wood of the elk or deer.
Dernabane (doireanach ban)—White woody place.
Drone (dron)—Straight hill or ridge.
Edintiloan (eadan toillin)—Hill face of the small hole or cave.
Glenroe (gleann ruadh)—Red glen.
Glack (glaic)—a round hollow or valley.
Garvey (garbh achadh)—Rough uncultivated land.
Golan (gabhlan)—A forked shaped glen or hill.
Innismagh (inis Macha)—Magha's holm or island; might also mean the island plain.
Killyneery (coille an aodhaire)—Wood of the shepherd.
Knockadreen (cnoc a draoighin)—Hill of the blackthorns.
Knocknarney (cnoc na n-airneadh)—Hill of the sloes.

Lisadavil (lios Dabhaill)—Dabhall's fort.
Lisbeg (lios beag)—Small fort.
Lisdoart (lios dubh Airt)—Art's black fort, or fort of the high mound.
Lisginny (lios gineadha)—Fort of the generation or conception (O'D.).
Loughans (lochan)—Small loughs.
Lisconduff (lios con duibh)—Fort of the black hound.
Leany (leanaidh)—A wet meadow.
Legaroe (loga ruadha)—Red hollows.
Mullaghbane (mullach ban)—White summit.
Mullaghnese (mullagh Neisi)—Enea's summit; or (mullach neasuighthe)—A hill coupled with another
Martray (martra)—A place where men were slaughtered.
Plaister (pleist-ire)—Stiff land.
Ravellea (rath bhile)—Fort of the large old tree.
Rousky (rusgaidh)—marshy or fenny land.
Reskatarriff (riasc a tairbh)—the bull's morass or quagmire.
Skey (sgiath)—A lone thorn bush.
Shantavney (sean tamhnadh)—Old green field, not cultivated for a long time.
Shanalurg (sean lurg)—Old track or path.
Tirelugan (tir an lagain)—District of dells or hollows.
Tullyvar (tulaigh bhfear)—Hill where men assembled for sports or other meetings.
Tullywinny (tul an mhuine)—Hill of the bracken or shrubbery.
Tully (tulaigh)—Gentle hills.
Tulnavern (tul na bhfearn)—Hill of the alders.

PARISH—CLONOE.

Annagher (eanach thoir)—Eastern marsh.
Annaghmore (eanach mor)—Great marsh.
Annaghnaboe (eanach na mbo)—Marsh of the cows.
Aughagalla (achadh a geala)—White fields; or (achadh gaille)—Field of the stone.
Aughagranna (achadh grandha)—Ugly field or bushy field.
Aghamullan (achadh Ui Maolain)—Mullan's field (O'D) (achadh a muilinn)—Field of the hill (Joyce).

Aughrimderg (achadh druim deirg)—Field of the red ridge.
Ballybeg (baile beag)—Little town.
Ballygittle (baile a n-giatail)—Town of the grain or of the rushes and reeds.
Clohog (clochog)—Stony land.
Cluntycracken (cluainte crocain)—Meadow of the spritty grass.
Clonoe (cluan eo)—Lawn of the yew trees.
Coole (cul)—The back of a hill.
Dernagh (doireanach)—Woody townland.
Drumurrer (druim muirir)—Ridge of the family or tribe.
Derryloughan (doire lochain)—Oak wood of the lake.
Derrytresk (doire triosca)—Grove of the brewer's grains, possibly the site of an old distillery.
Gortnagloch (gort na g-cloch)—Stony field.
Killary (coill an riogh)—King's wood or (caol shaire) red narrow inlet.
Killeen (cillin)—Little church.
Lisaclare (lios an chlair)—Fort of the level plain.
Lisnastrane (lios an tsruthan)—Fort of the streamlet or burn.
Lendadremnagh (leana driumneach)—Meadow of the ridges.
Magheramulkenny (machaire muil chinn)—Plain of the prominent or bare hill.
Meenagh (meadhanach)—Central land, between two other townlands.
Magheralamfield (machaire leamhcoille)—Plain of the elm wood.
Shanliss (sean lios)—Old fort.
Tumpher (tom feir)—Place where bunchy grass grows.

Lisadavil (lios Dabhaill)—Dabhall's fort.
Lisbeg (lios beag)—Small fort.
Lisdoart (lios dubh Airt)—Art's black fort, or fort of the high mound.
Lisginny (lios gineadha)—Fort of the generation or conception (O'D.).
Loughans (lochan)—Small loughs.
Lisconduff (lios con duibh)—Fort of the black hound.
Leany (leanaidh)—A wet meadow.
Legaroe (loga ruadha)—Red hollows.
Mullaghbane (mullach ban)—White summit.
Mullaghnese (mullagh Neisi)—Enea's summit; or (mullach neasuighthe)—A hill coupled with another
Martray (martra)—A place where men were slaughtered.
Plaister (pleist-ire)—Stiff land.
Ravellea (rath bhile)—Fort of the large old tree.
Rousky (rusgaidh)—marshy or fenny land.
Reskatarriff (riasc a tairbh)—the bull's morass or quagmire.
Skey (sgiath)—A lone thorn bush.
Shantavney (sean tamhnadh)—Old green field, not cultivated for a long time.
Shanalurg (sean lurg)—Old track or path.
Tirelugan (tir an lagain)—District of dells or hollows.
Tullyvar (tulaigh bhfear)—Hill where men assembled for sports or other meetings.
Tullywinny (tul an mhuine)—Hill of the bracken or shrubbery.
Tully (tulaigh)—Gentle hills.
Tulnavern (tul na bhfearn)—Hill of the alders.

PARISH—CLONOE.

Annagher (eanach thoir)—Eastern marsh.
Annaghmore (eanach mor)—Great marsh.
Annaghnaboe (eanach na mbo)—Marsh of the cows.
Aughagalla (achadh a geala)—White fields; or (achadh gaille)—Field of the stone.
Aughagranna (achadh grandha)—Ugly field or bushy field.
Aghamullan (achadh Ui Maolain)—Mullan's field (O'D) (achadh a muilinn)—Field of the hill (Joyce).

Aughrimderg (achadh druim deirg)—Field of the red ridge.
Ballybeg (baile beag)—Little town.
Ballygittle (baile a n-giatail)—Town of the grain or of the rushes and reeds.
Clohog (clochog)—Stony land.
Cluntycracken (cluainte crocain)—Meadow of the spritty grass.
Clonoe (cluan eo)—Lawn of the yew trees.
Coole (cul)—The back of a hill.
Dernagh (doireanach)—Woody townland.
Drumurrer (druim muirir)—Ridge of the family or tribe.
Derryloughan (doire lochain)—Oak wood of the lake.
Derrytresk (doire triosca)—Grove of the brewer's grains, possibly the site of an old distillery.
Gortnagloch (gort na g-cloch)—Stony field.
Killary (coill an riogh)—King's wood or (caol shaire) red narrow inlet.
Killeen (cillin)—Little church.
Lisaclare (lios an chlair)—Fort of the level plain.
Lisnastrane (lios an tsruthan)—Fort of the streamlet or burn.
Londadremnagh (leana driumneach)—Meadow of the ridges.
Magheramulkenny (machaire muil chinn)—Plain of the prominent or bare hill.
Meenagh (meadhanach)—Central land, between two other townlands.
Magheralamfield (machaire leamhcoille)—Plain of the elm wood.
Shanliss (sean lios)—Old fort.
Tumpher (tom feir)—Place where bunchy grass grows.

OMAGH UNION.

PARISH—DROMORE.

Aghadulla (achadh dealbh)—Field of the phantoms or ghosts (Joyce).
    (achadh tullach)—Field of the summit (O'D)
Aghadarragh (achadh darach)—Field of the oak wood.
Aghee (achadh aoi)—Field of the shepherds.
Aghlisk (each laisc)—An enclosure for horses.
Aghnamoe (achadh na m-bo)—Field of the cows.
Bodoney (both Domhnaigh)—Sunday hut, set apart for religious worship.
Camderry (cam doire)—Crooked oak wood or grove.
Corrasheskin (curragh an t-seiscinn)—Soft place of the quagmire.
Curly (cor liath)—Round grey hill.
Carnalea (carn liath)—Grey cairn or stone heap; or (carn na laogh) cairn of the calves.
Corbally (cor baile)—Odd townland; or
    (cuar baile)—Round shaped townland.
Corlaghdergan (cor loch Ui Deargain)—Round hill of Dergan's lough.
Cornamuck (cor na muc)—Hill of the pigs.
Cornamucklogh (cor na muclach)—Hill of the piggeries
Coyagh (caidheach)—Full of natural trenches.
Cranny (crannach)—A place abounding in bushes.
Dergany (dearg eanach)—A red marsh.
Doocroch (dubh chnoc)—A black hill.
Dressoge (dreasog)—Brambly land.
Drumlish (druim leasa)—Ridge of the fort.
Drumsheil (druim Saidheil)—Sheil's ridge (O'D.).
    (druim saileoga)—Ridge of the willows (Joyce).
Dromore (druim mor)—The big ridge.
Drumconnis (druim conaidh)—Ridge where firewood is got.
Drumskinny (druim scine)—Knife shaped ridge.
Derrynaseer (doire na saor)—Wood or hill of the carpenters or mechanics.
Dullaghan (tulchan)—A small hill (O'D.).
    A place of ghosts where a headless man appears (Joyce).
Drumderg (druim dearg)—Red ridge.

Drummallard (druim mala aird)—Ridge of the high or overhanging brow.

Esker (eiscir)—A low gravelly ridge.

Edenagon (eadan na g-con)—Hillface of the hounds.

Galbally (gall-baile)—Foreigner's or English town.

Golan (gabhal-an)—Fort shaped portion of land.

Gardrum (gearr dhruim)—Short ridge.

Greenan (grianan)—A sunny situation.

Glengeen (gleann gaoine)—Glen of goodness or beautiful glen.

Knockaraven (cnoc)—Raven's hill.

Kildrum (coill droma)—Ridge of the wood.

Knocknahorn (cnoc na h-eorna)—Barley hill.

Lettergash (litir gaise)—Swift water current of the wet hillside.

Lettery (leithreach)—A wet or "spewy" hillside.

Lissaneden (lios an eadain)—Fort of the hill brow.

Mullaghbane (mullach ban)—White summit.

Mullanboy (mullan buidhe)—Yellow summit.

Magheragart (machaire gairt)—A renowned or famous plain.

Meenagowan (min a ghabhann)—The smith's smooth field.

Mulnagoagh (mulna g-cuach)—Hill of the cuckoos.

Meenagar (mine gearr)—Short mountain meadow.

Oughterard (uachtar ard)—Upper height or summit.

Polfore (poll fuar)—Cold pit or hole.

Rahony (rath an chonnaidh)—Fort of the firewood.

Rakeeranbeg (rath caorthain bhig)—Rath of the little quicken tree grove.

Shannaragh (sean rath)—Old rath or fort.

Straduff (srath dubh)—Black holm or strand.

Shanmullach (sean mhullach)—Old summit or hilltop.

Skeogue (sciathog)—A lone thorn bush.

Tullymagough (tulaigh Mhic Eochadha)—Keogh's hill.

Tullywee (tulaigh bhuidhe)—Yellow hill.

Tummery (an t-iomaire)—The ridge or raised land.

Tattyolunagh (tate cluaine)—Division of the meadows.

Tattyoor (tate cor)—Land division of the cranes; or land division of the round hill.

PARISH—LONGFIELD EAST.

Carony (cor Una)—Una's or Ethna's hillock.
Claraghmore (claragh mor)—Large bare level plain.
Cornavarrow (cor na bhfaraid)—Round hill or platform of the meeting place.
Coolkeeragh (cul caorach)—Hill back where sheep graze.
Drumbarley (druim barr liath)—The grey topped ridge.
Drumhonish (druim Chonais)—Conish's ridge; or
      (druim con uisge)—Ridge of the waterdog.
Drumnaforbe (druim na foirbe)—Ridge of the land.
Drumrawn (druim ramhain)—Spade shaped ridge.
Drumquin (druim caoin)—Pleasant ridge.
Dressogue (dreasog)—A brambly place.
Gortaghar (gort eachraidh)—Field of the cattle fold.
Garvaghullion (garbh achadh cuilinn)—Rough holly growing land.
Leganvy (lag an bheithe)—Pool around which birch trees grow.
Legphressy (lag preasach)—A bushy pool or hollow.
Laght (leacht)—A monument or monumental stone.
Magharenny (machaire an etnaigh)—Marshy plain.
Segully (sidhe Guile)—Guil's fairy hill (O'D.).
      (guail)—A place where coal or charcoal is got (Joyce).
Unshinagh (fuinnseannach)—Abounding in ash trees.

PARISH—CAPPAGH.

Aghalane (achadh leathan)—A broad field.
Aghagallon (achadh Gallon)—Gallon's field or the field of the standing stones.
Arvalee (airbhe laoigh)—Land division on which calves are reared.
Beltany (baile teine)—Town of the druidic fires which were lit on May eve.
Beragh (beith ratha)—Rath or fort of the birch tree.
Ballykeel (baile caol)—Narrow townland.
Ballynatubbrit (baile na tíobraide)—Town of the well or fountain.
Ballynamullan (beal atha na muilleann)—Ford of the mills.
Boheragh (bothar ratha)—Road of the fort.

Bunnynubber (bun an ubhar)—Bottom land of the mire or marsh.
Ballynaguilly (baile na cuile)—The corner town; or (baile na coille) town of the wood.
Cloughfin (cloch fhionn)—A place of white (quartz) stones.
Carrigans (carraigan)—A place of little rocks.
Castleroddy (caislean Rodaigh)—Roddy or Roger's castle.
Cappagh (ceapach)—Plot of land laid out for tillage.
Cullion (cuileann)—A place where holly grows.
Cranny (cranaidh)—Arboreous or place full of bushes or trees.
Crosh (cros)—A cross.
Carnony (cor an chonaidh)—Round hill of the bushwood.
Corranarry (cor an aedhaire)—The shepherd's hill.
Calkill (coll choill)—A hazel wood.
Creevenagh (craobhanach)—A branchy place.
Conywarren (coinegear)—A rabbit warren.
Campsie (camasaigh)— a bend in the river, or a winding road.
Camowen (cam abhainn)—Crooked river.
Dunbreen (dun Braoin)—Breen's fort.
Dunmullan (dun mullan)—Fort of the little hill.
Dergmoney (dearg muine)—Red shrubbery.
Eskeradooey (eiscir a dubhaigh)—Sand ridge of the black dye stuff.
Erganagh (earganach)—Good fertile land.
Edenderry (eadan doire)—Hill face of the oak wood.
Farrest (foraois)——A forest (likely an English word)
Fernagh (fearnach)—Abounding in alder trees.
Faccary (faithche cora)—Green or wood of the weir.
Gortgranagh (gort granach)—Field of the grain.
Gortinagin (gort na g-cinn)—Little field of the heads.
Glencordial (gleann Chordial)—Cordial's glen.
Glengawna (gleann gabhna)—Glen of the smiths.
Golan (gabhlan)—Fork shaped glen or hill.
Gortnacreagh (gort na creiche)—Field of the plunder.
Galbally (gall bhaile)—Foreigner's or English town.
Garvaghy (garbh achadh)—Rough uncultivated field.
Innisglan (inis gleann)—Strong stream of water proceeding from an island shaped glen.

Killinure (coill an uibhair)—Yew tree wood.
Kingarrow (ceann garbh)—Rough headland or hillside.
Killyclogher (coill an chlochair)—Wood of the stone structure or church, or of stoney field.
Killybrack (coill bhreac)—Speckled wood.
Knockmoyle (cnoc maol)—Bald or bare hill.
Killycurragh (coill an churraigh)—Marshy wood.
Killins (coillin)—Small woods or churches.
Legacurry (lag a choire)—Hollow of the cauldron or big pot.
Lislap (lios na leabhtha)—Fort of the bed or grave.
Lurganbuoy (lurga buidhe)—Long yellow hill.
Lisnaharney (lios na h-airneagh)—Fort of the sloes.
Lisnagirr (lios na g-caor)—Fort of the berries (O'D.).
　　　　　(lios na g-carr)—Fort of the cars (Joyce).
Lisnamallard (lios a mala ard)—Fort on the high side of a hill.
Lisahoppin (lios a chopain)—Cup shaped fort.
Lisboy (lios buidhe)—Yellow fort.
Lislea (lios laith)—Grey fort.
Maine (meadhan)—Middle town or little plain.
Mullaghmore (mullach mor)—The big summit.
Omagh (omhaigh)—A complete or sacred plain ("o" in old Irish signifies sacred).
Reaghan (riachan)—Small grey spot of land.
Rossnamuck (ros na muc)—The plain or wood of the pigs.
Rylagh (raileach)—Abounding in oak.
Racolpa (rath colpach)—Fort or rath of the heifers.
Recarson (reidh Carson)—Carson's flat land.
Straughroy (srath ruadh)—Red river holm; or
　　　　　(srath riogh)—The King's holm.
Tattynure (tate an iubhair)—Land division where yew trees grow.
Tirour (tir corr)—District of the cranes.
Tirmurty (tir Murtagh)—Murty or Murtagh's district.
Tirquin (tir O'Chionn)—Quinn's district.
Tantramurry (tan srath murbhach)—Flat marshy region along a river bed.
Tattynagole (tait na g-coll)—Land division of the hazels
Tattraconnaghty (tait na Chonnachtaigh)—Connaughtman's rath or division of land.
Tully (tulaigh)—A hill or gentle slope.

PARISH—DRUMRAGH.

Aghaleag (aghadh na lice)—Field of the large flat stone

Aghnamoyle (achadh na macile)—Field of the hornless cow.

Aghadulla (achadh a dealbha)—Field of the phantom or ghost.

Aghoe (achadh aoi)—Field of the island.

Ballygowan (baile gabhann)—Blacksmith's town.

Botera (both aodhaire)—A shepherd's hut.

Ballynahatty (baile na thaite)—Town of the tate or land division; or it might mean (haithe) town of the lime kiln. (A tate is about 60 acres.)

Beagh (beigheach)—A birch wood.

Blacksessiagh (seiseadach)—Black sixth part.

Clohog (clochog)—A stony place.

Corlea (cor liath)—Grey hill.

Culbuck (cul na m-boc)—Hill back of the bucks.

Cavanacaw (cabhan a' chatha)—The round hill of the chaff.

Clanbogan (cluan Bogan)—Lawn of the sort surface.

Corradinna (cor a diona)—Hill of the small fortress; or (cor a dinne) hill of the man (possibly some noted person).

Creevanger (craobhan gearr)—Low bushes.

Creevanmore (craobhan mor)—Large bushes.

Crucknamona (cnoc na mona)—Hill of the bogs or moss

Coolnagard (cul na g-ceard)—Hillback of the artificers or tradesmen.

Cornabracken (cor na mbreacan)—Hill of the speckled land.

Culmore (cul mor)—Big hill back.

Dunwish (dun ghuise)—Fort or rath of the fir trees.

Drumconnolly (druim Connolly)—Connolly's ridge.

Drudgeon (druidgeon)—An enclosure.

Drum (druim)—A ridge or hill back.

Drumragh (druim an ratha)—Ridge of the rath or fort.

Drumshanley (druim an cean tighe)—Ridge of the old house (Joyce); (druim h-Ainlighe)—Hanley's ridge (O'D.).

Dergmony (dearg mhuine)—Red shrubbery.

Doogary (dubh charaidh)—Black weir.

Edergoole (idir gabhail)—Land enclosed by two branches of a river.

Fireagh (fiodh riach)—A grey wood.
Freughmore (fraoch mor)—The great heath.
Gillygooly (coill na cuile)—Corner wood.
Garvagh (garbh achadh)—Rough ground or field.
Gortrush (gort an ruis)—Field of the wood.
Gortmore (gort mor)—Large field.
Gammy (geamach)—Cold, exposed, winter like townland.
Kilmore (coill mhor)—The great wood.
Kiltamnagh (coill tamhnach)—Wood of the green fields
Kivlin (coibhlin)—A place where long tufts of grass grow.
Lissan (liosan)—A little fort.
Loughmuck (loch muc)—Lough of the pigs.
Lammy (leamhach)—A place of elms.
Mullagharn (mullach a chairn)—Summit of the stone heap.
Mullanatoomog (mullach na tumoige)—Summit of the burial mound.
Mullaghmore (mullaghmor)—Big summit
Mullaghmenagh (mullach meadhanac)—Middle summit
Rylands (reidhlean)—Green field for dancing or amusement.
Relaghdooey (reidh or relagh dubh)—Black level field.
Rakeeragh (rath caorach)—Fort of the sheep.
Stroancarbadagh (sruthan carbadach)—Streamlet of the chariot.
Sedennan (suidhe daingean)—Set of the fortress or stronghold.
Tarlum (tur or tor lom)—Bare tower or bush.
Tattyreagh (tate riabhach)—Grey tate or division.
Tamlaght (tamlacht)—A burial mound.
Tattykeel (tate caol)—Narrow division.
Tullycunny (tulaigh conaigh)—Hill of the firewood.

PARISH—CLOGHERNY.

Aghnagar (achadh na g-corr)—Field of the herons.
Annagh (eanach)—A marshy place.
Ballyhallaghan (baile Ui Ealachain)—O'Hallaghan's town.
Ballykeel (baile caol)—Narrow town.
Beragh (bearnach)—A gapped place; or (beitha) field of the birches.

Clogherny (clochernaigh)—A stony place (Joyce); or (cloch airne)—A stony place where sloes grow (O'D.).
Coolesker (cul eiscir)—Back of the low gravelly hills.
Curr (cor)—A round hill.
Donaghanie (domhnach an eich)—Church of the steeds
Dorvaghroy (derbhagh riabhac)—Gray grove.
Eskermore (eiscir mor)—Big esker or gravelly hill.
Gortaclare (gort a chlair)—Field of the leven plain.
Kilcam (coill cham)—A crooked wood.
Killadroy (coill a druadh)—The druid's wood.
Laragh (laithreach)—Site or situation of a building.
Legacurry (lag a choire)—Pool of the cauldron or marsh.
Letfern (leacht fearn)—Monument of the alder trees.
Moylagh (maol achadh)—Bare or bald field or hill.
Mullaghmore (mullach mor)—The big summit.
Ranelly (rath Neillidh)—Nelly's fort; Nelly was a man's name.
Rarone (rath ruadhan)—Reddish fort.
Radergan (rath Deargain)—Dargan's fort.
Raw (rath)—A fort.
Roscavey (rus ceibhe)—Wood or point of the long grass.
Seskinore (sheskin odhar)—Pale grey marshy bog.
Tulyheerin (tulaigh chaorthainn)—Hill of the quicken tree.
Tullyrush (tulaigh ruis)—Hill of the point or wood.
Tattykeerin (tate caerthainn)—Land division of the rowan tree.

PARiSH—LOWER BADONEY.

Aghascrebagh (achadh scriobach)—Furrowed or rugged land. There is a celebrated Ogham stone here, hence some say it may mean the field of the writing
Aloolies (aill bhuaille)—Dairy of the spink or glen.
Altacamcossy (alt na camcoise)—Crooked leg shaped glen.
Attagh (ait toighe)—Site of a house.
Aghaboy (achadh buidhe)—Yellow field.
Aghnamerrigan (achadh na muireagan)—Field of the foxgloves.
Badoney (both Domhnaigh)—Sunday hut where prayers are said.

Brackagh (breacach)—Speckled land.
Beltrim (beal tirim)—Dry river mouth (O'D.).
　　　(beal truim)—Ford mouth of the alder trees (Joyce).
Binnafreaghan (binn a phreacain)—Headland of the ravenous bird.
　　Joyce gives it " peak of the whortleberry.
Casorna (cuas eorna)—Cave of the barley field.
Crockanbuoy (cnocan buidhe)—Little yellow hill.
Curraghanalt (curragh an ailt)—The moor of the glen.
Cashel (caiseal)—A circular stone fort.
Carnanrancy (carn an raithin sidhe)—Carn of the fairy rath or ferns.
Crock (cnoc)—A hill.
Culvahullion (cul mhaighe cuilinn)—Back of the holly plain.
Drumlea (druim liath)—Grey ridge.
Droit (droichead)—A bridge.
Dunbunrawer (dun bun reamhar)—Fort of the thick bottom or hill base.
Formal (for maol)—Bare or bald hill.
Fallagh (faill ath)—Fort on the spink.
Glenlark (gleann leirge)—Glen of the hill slope.
Greenan (grianan)—A sunny situation.
Glenmacoffer (gleann Mhic Cathbhairr)—M'Caffer's glen. He was found dead there.
Gortin (gortin)—A little field.
Garvagh (garbh achadh)—Rough field or land.
Gorticashel (gort an chaisil)—Field of the circular stone fort.
Keerin (caertheainn)—Boggy land on which the mountain ash grows, or a moor.
Leaghan (leathan)—A wide or broad piece of land.
Lenagh (leuna)—Wet meadowy land.
Leggatraght (lag a t-sneaghta)—The hollow of the snow; or (lag a t-sratha) hollow of the gravelly holm (O'D.).
Leggins (leagain)—Little hollows.
Lenamore (leuna mor)—Big wet meadow.
Liscabble (lios capall)—Fort where horses were enclosed at night.
Monanameal (muine na miol)—Bog of the flies.

Meenarodda (min a roda)—Field of the ferruginous scum, known as spa (red coloured) water.

Meenadoo (min a dubh)—Black rough meadow land.

Oughminacroy (ucht mine cruaidh)—Breast or side of the hard mountain meadow.

Rouskey (rusgaidh)—Marshy land or coarse bent grass.

Rylands (reidhlean)—Level green field used for sports, or games. O'Donovan says the meaning is uncertain

Sheskinshule (sescein suibhail)—Moving bog grass; long grass waving in the wind.

Stradawan (srath domhain)—Deep holm, or holm of the oxen.

Teebane (taobh ban)—Sunny or bright side, having a southern aspect (O'D.).
    (tigh ban)—White house (Joyce).

Tievebrack (taobh breac)—Speckled hillside.

Trinamadan (trian an amadain)—The fool's third part or divsion.

Tievenameena (taobh na mine)—Side of the mountain meadow.

PARISH—DONACAVEY.

Agharonan (achadh Ronain)—Ronan's field (Joyce). Reddish field (O'D.).

Aghafad (achadh fada)—Long field.

Attaghmore (ait tighe moir)—Ste of the big house.

Annaghbo (eanach bo)—Marsh of the cows.

Annaghmurrin (eanach moirin)—Coarse, marshy grass land.

Ardatinny (ard na teine)—Hill of the fires.

Belnagarnan (beal na g-carnan)—Mouth of the stone heaps or carns.

Baronagh (baranach)—Top lands or headlands.

Carryglas (carraig ghlas)—Grey rock.

Corrashesk (currach seisge)—Moor of the sedge or quagmire.

Crocknafarbrague (cnoc na bhfear breige)—Hill of the false or lying man.

Carnorousg (cari a ruisg)—Carn of moor or morass.

Cavan (cabhan)—Means a hill in the North, in other parts of Ireland a hollow.

Corbally (cor bhaile)—Odd townland or hilly townland

Cranny (cranaidh)—Arborous or place of trees.

Carnalea (cor na laogh)—Hill of the calves.
Cumber (cumar)—A confluence or meeting of waters.
Derrybard (doire baird)—Bard's oak grove.
Draughton (doire Fhacthain)—Fagton's oak wood.
Drumlagher (druim lathair)—Ridge of the site of a house; also a battlefield.
Dundivin (dun Daimhin)—Devin's fort.
Dungoran (dun garrain)—Fort of the shrubbery.
Donacavey (Domhnach Cheibhe)—Cavey's church.
Drumwhisker (druim chosgair)—Ridge of victory, where a battle was fought.
Dunnamona (dun na mona)—Fort of the bog.
Drummond (dromann)—A long ridge or hill.
Edenafogry (eadan a fograidh)—Brow of the shelving hillside; or hill face of the warning (O'D.).
Edenatoodry (eadan a t-sudaire)—Hill face of the tanner.
Edenasop (eadan na sop)—Hillface of the wisps.
Fallaghearn (falach fhearn)—Hedges of alder trees.
Feenan (fionan)—White land.
Fintona (fionn tamhnach)—White field or land.
Freughmore (fraoch mor)—Big heath.
Garvallach (garbh allach)—Rough land.
Gargrim (gearr dhruim)—Short ridge.
Glennan (gleannan)—Small glen or valley.
Gulladoo (gualla dubh)—Black mountain shoulders.
Killymoon (coill Ui Mhonain)—Moonan's wood.
Killyliss (coill an leasa)—Wood of the fort.
Kilcootry (coill cuit riabhaigh)—Wood of the wild grey cat.
Kilgort (coill gort)—The wooded field.
Kilberry (coill Ui Bhearaigh)—Berry's wood.
Lisconcrea (lios con riabhaigh)—Fort of the brindled hound.
Lurganbuoy (lurgan buidhe)—Long yellow hill.
Lackagh (leaca)—A stony hillside.
Legamaghery (lag machaire)—Hollow of the plain.
Lisnacrieve (lios na craoibhe)—Fort of the spreading tree.
Legatiggle (lag a t-seagail)—Hollow or valley of the rye.
Lisavaddy (lios a mhadaidh)—Dog's fort.

Lisdergan (lios Deargain)—Dergan's fort or red coloured fort.
Lisky (lios sceithe)—Fort of the bushes; or (lios sciath) fort of the shields (Joyce).
Lisnabulrevy (lios na b-poll riabhach)—Fort of the grey holes or pools, or borders.
Lisnagardy (lios na g-ceardchan)—Fort of the forges or workshops.
Mullanboy (mullan buidhe)—Yellow little hill.
Mullawinny (mullach mhuine)—Summit of the brake or shrubbery.
Mullans (mullan)—Little hills.
Mullasiloga (mullach na saileoga)—Hill of the willows.
Raneese (rath Aonghusa)—Angus's fort.
Racrane (rath Cridhean)—Crean's fort or rath of hard clay.
Rathwarren (rath Bharrain)—Baron's fort or rath of the rabbit burrow or gap.
Roughan (ruadhchan)—Reddish land.
Rarone (rath ruadhan)—Red rath or fort.
Strabane (srath ban)—White strand or holm.
Sessiagh (seiseadhach)—Sixth part or division.
Skreen (scrin)—A shrine.
Stratigore (srath toigh gabhair)—Holm of the goat house.
Syonfin (siodhan fionn)—White fairy hill.
Screggagh (screagach)—Rocky land covered with shrubbery.
Stranisk (srath an uisge)—Holm often submerged by floods.
Syonee (siodh an Aodha)—Hugh's fairy hill.
Skelgagh (sgeilgeach)—Rocky land.
Tullyrush (tul an ruis)—Wood of the little hill.
Tullyvalley (tul an bhealaigh)—Hill of the way, pass or road.
Tireenan (tir grianain)—District of the sunny situation
Tonnagh Beg (tamhnach beag)—Little green field.
Tonnagh Mor (tamhnach mor)—Large green field.

PARISH—ARDSTRAW.

Altadoghal (alt an diochill)—Glen of the great wood, or (alt Dubhghcoill) Dowell's glen or spink.
Aghafad (achadh fada)—Long field.

Aghasessy (achadh seasmach)—The standing or reliable productive field, part or division.

Ardstraw (ard srath)—High holm or height of stress, effort or difficulty.

Ardbarren (ard barran)—The high gap.

Bloomry—O'Donovan says it seems an English word.

Ballought (baile leachta)—Town of the monuments.

Breen (bruighean)—A fairy palace.

Ballynaloan (baile na lon)—Town of the light, brightness or radiance.

Ballyfolliard (baile Folliard)—Folliard's town; or (baile faille aird (e) town of the high spink.

Ballyrenan (baile Ui Roonain)—Renehan's town; or town of the ferny shrubbery.

Birnaghs (bearnach)—A gapped place.

Brocklis (broch lios)—Fort of the badgers; or a den, cave or vault.

Ballymullarty (baile Maol Fheartaigh)—Mullarty's town, or town of the high summit.

Beagh (beithe)—Land abounding in birches.

Bunderg (bun dearg)—Red bottom land, or mouth of the R. Derg.

Binnawooda (binn a mhuada)—The cloudy peak.

Bolaght (both leacht)—The hut of the monument; or hut for cattle or a dairy place.

Carncorran (carn corran)—A hook shaped heap of stones.

Carrickadartan (carraig a'dartain)—Rock of the herd or rock of the small projection.

Castlebane (caislean ban)—White castle.

Coolnacrunaght (cul na cruithneacht(a)—Hill back of the wheat.

Claremore (clar mor)—Great plain.

Coolreaghy (cul riabhach)—Grey hill back unfenced.

Coolnaherin (cul na h-ireann)—Back of the field or land containing iron ore.

Crew (craobh)——A branching tree.

Cavandarragh (cabhan darach)—Hill of the oak wood.

Creevy (craiobheach)—Bushy land.

Carnaveagh (carn an bheithe)—Carn of the birches.

Crosh (cros)—A cross.

Croshballinree (cros baile 'n riogh)—Cross town of the king.

Clady (claidhighe)—Muddy margin of a stream or river.
Casty (casta)—Place of the entangled grass or shrubs.
Cloonty (cluainte)—Meadowy land.
Carnkenny (carn caonaigh)—Mossy carn.
Coolaghy (cul achadh)—Back of the field or hill side or a vault.
Cloughogle (cloch thogbhalta)—A raised or lifted stone.
Drumnabeg (druim na mbeithe)—Ridge of the birches.
Dunrevan (dun Reabhain)—Reavan's dun or rath.
Derrygoon (doire gamhain)—Wood of the calves; or (gabhann) wood of the smith (s), or a shrunken wood.
Drumclamph (druim clam)—Ridge of the lepers.
Dunteige (dun Teige)—Teige or Timothy's dun or rath
Drumlegagh (druim liagac)—Stony ridge.
Drumnahoe (druim na h-uamha)—Ridge of the cave.
Erganagh (earganach)—Good fertile land.
Envagh (einmheach)—Rich or abundant place.
Fyfin (faithche fionn)—Fair green fields.
Garvetagh (garbh aileamh)—Rough land.
Golan (gabhlan)—Fork-shaped land.
Glenglush (gleann glaise)—Glen of the streamlet.
Glasmullagh (glas mullach)—Green summit.
Gallan (gallan)—A place where standing stones are found.
Glenknock (gleann cnuic)—Glen of the hill.
Killeen (cilllin)—A little church; or (coillini) little woods.
Knockiniller (cnoc an iolair)—The eagle's hill.
Kilreal (cill Redhgeal)—Church of St. Regulus.
Kilstrule (coill sruille)—Wood of the stream.
Knockbrack (cnoc breac)—Speckled hill.
Knockroe (cnoc ruadh)—Red hill.
Kilymore (coille mhor)—Big wood.
Killydart (coille dart)—Wood of the herd.
Listymore (lios toigh' moir)—Fort of the big house.
Lurganboy (lurga buidhe)—Long yellow hill.
Lisnacreaght (lios na gcreachta)—Fort of the raids or of the ravine.
Legnabraid (lag na braghaid)—Hollow of the gorge or deep cutting.

Lisnafin (lios na finne)—Fort of the white animal or thing.
Lisnatunny (lios na tonnaighe)—Fort of the quagmire (lios na teine)—Fort of the bonfires on 23rd June (Joyce).
Liscreevaghan (lios craobhachain)—Fort of the branching trees.
Largybeg (leargaidh beag)—Small sloping land.
Legland (leith ghleann)—Half glen or small valley.
Letterbin (leitir binn)—Soft watery peaked hill.
Laragh (laithreach)—Site or situation of a building.
Ligfordrum (lag fuar dhruim)—Pool of the cold summit from which a good view is obtained.
Lettercarn (leitir carn)—Wet spewy carn or stone heap
Lisleen (lios lin)—Fort of the flax or gray coloured fort
Magheracreggan (machaire creagan)—Plain of the little rocks.
Magheralough (machaire locha)—Plain of the lough.
Moyle (maol)—A bald or bare place.
Mulvin (mull an bhinn)—Hill of the spink.
Mullach (mullach)—A hill top.
Magheracolton (machaire Colton)—Colton's plain, or the sylvan plain.
Meaghy (machaidh)—Cattle fields. It has as antiquities the Giant's Den and four Danish forts.
Priestsessiagh (seiseadhach)—Priest's sixth part of division.
Pubble (pobal)—A congregation; an open-air place of assembly; also a tent or pavilion.
Ratyn (rath terne)—Fort or hill on which fires were lighted; or ivy rath.
Rakelly (rath Ceallaigh)—Kelly's rath or dun, or narrow rath.
Shanog (seanog)—Little hill of the foxes.
Scarvagherin (scarbhach iarainn)—Gravelly place containing iron ore.
Sessiagh (seseadhach)—A sixth part or division.
Shanonny (sean thonaich)—Old mound (O'D.). (sean dhonaigh)—Old church (Joyce).
Strahulter (srath ultar)—Holm of the plague graveyard; or a delightful holm cultivated with the plough.

Straletterdallan (srath leitir dallan)—Wet holm of the winnowing; or of the wet hill of the leeches.

Stonyfalls (toin a' falaidh)—Enclosed bottom land.

Skinboy (ais cinn buidhe)—Side of the yellow hill.

Tullymuck (tullach muc)—Hill of the pigs.

Tamnagh (tamhnach)—A green field.

Tievenny (taobh eanaigh)—Marshy hillside of the hanging down grass.

Tirmegan (tir Megan)—Megan's district.

PARISH—TERMONMAGUIRK.

Aghanereagh (achadh na reagha)—Little field of the sorrow or pleasant field.

Altanagh (altan ach)—Abounding in cliffs or glens, or steep glen of the fort.

Athenree (achadh an riogh)—King's field or fort (O'D.). Little fort of the heath (Joyce).

Aghagogan (achadh goigin)—Field where giddy headed people congregate; or field of the cackling.

Aghnagreggan (achadh na g-creagan)—Field of the little rocks.

Altdrumman (alt droman)—High hill ridge.

Aghnaglea (achadh na gcliath)—Ford or field of the hurdles.

Bancran (beann crann)—Peak of the trees.

Bracky (breacaidhe)—Speckled land.

Ballintrain (baile an trein)—Town of the brave or mighty man; or town of the murmuring wind.

Carrickmore (carriag mhor)—Big rock.

Clare (clar)—A level plain.

Creggandevesky (creagain dubh uisge)—Rock of the black or mossy water.

Creggan (creagan)—A rocky place.

Cregganconroe (creagan con ruaidh)—Little rock of the red hound (fox).

Copney (copnaidh)—Abounding in dock leaves.

Cavanreagh (cabhan riabhac)—Grey hill.

Cloghfin (cloch fhionn)—A place of white (quartz) stones.

Cooley (cuailne)—Place of stakes or faggots.

Dunmisk (dun mis)—Dun of distress or poverty.

Drumnakilly (druim na coille)—Ridge of the wood.

Deroran (doire Oran's)—Oran's wood; or wood of the cream-coloured horse.

Derroar (doire odhar)—Grey or brown oak wood.
Drumduff (druim dubh)—Black ridge.
Drumlester (druim lestar)—Ridge of the vessels.
Eskerboy (eiscir buidhe)—Yellow esker or low gravelly hill.
Granagh (greanach)—A gravelly place.
Gleneeny (gleann eidhnean)—Ivy producing glen; or (gleann einigh) hospitable glen.
Gortfin (gort fionn)—Fair or white field.
Gortfinbar (gort Finbar)—Finbar's field.
Innishatieve (innis a taoibh)—Side of the river holm.
Loughmacrory (loch MacRuairdhi)—M'Rory's lough.
Liscincon (lios cinn con)—Fort of the hound's head.
Mullan (mullan)—A little hill.
Mulnafye (mul na faithche)—Hill of the sporting green
Ramackan (rath meacan)—Field of the parsnips.
Sluggan (slogan)—A hole in a river where the water is slugged or sucked underground.
Sultan (sailtean)—A place where willows grow, or a water cut channel.
Skeboy (sgiath buidhe)—Yellow bush or thorn.
Streefe Glebe (sraobh or srae)—A mill stream.
Tiroony (tir uamhnaighe)—District of the terror or fear.
Tonegan (tonagan)—Boggy bottom lands.
Tremoge (tromog)—Place of alder trees.
Tursallagh (tur salach)—Miry hill.
Termon Rock (termon)—Church land.
Tanderagee (toin le gaoith)—Town with its "backside" to the wind.

STRABANE UNION.

PARISH—DONAGHEDY.

Ardcame (ard cam)—The crooked height.
Ardmore (ard mor)—The great height.
Aughtermoy (uuchter maigh)—The upper plain.
Altrest (alt trist)—Glen of the curse.
Aghafad (achadh fada)—The long field.
Altishane (alt a t-siodhain)—Glen of the fairy fort.
Aghabrack (achadh breac)—The speckled field.
Ballynacross (baile na croise)—Town of the cross.
Ballyneaner (baile n-eanar)—Lonely town.
Bunowen (bun abhann)—Mouth of the river.
Ballybeeny (baile beanaigh)—Reaping townland; or (baile beanaimh)—Town good for yielding grain.
Ballyheather (baile iochtair)—Lower townland.
Ballaghlare (bealach lair)—Middle road.
Barran (barran)—A small hill or summit.
Binnelly (binn aille)—The peak of the "spink."
Ballykeery (baile caerthainn)—Town of the rowan tree
Ballynabwee (baile an atha buidhe)—Town or mouth of the yellow ford.
Ballynamallaght (baile na mallacht)—Town of the curses.
Binbunnif (binn bun duibh)—Peak of the black bottom.
Balix (bealaigh)—Roads or passes.
Carnagribban (carn na criadh baine)—Carn of the clay marl; or (carn na Ghricbain)—Gribbon's carn.
Claggan (cloigean)—Round rocky hill or bare headland
Creaghan (criochan)—A grove or shrubbery (O'D.); or a little boundary (Joyce).
Cloghboy (cloch bhuidhe)—Yellow stone.
Cloghogle (cloch thogalta)—Raised or lifted stone.
Coolmaghery (cul machaire)—Back of the plain.
Creaghcor (creach cor)—The brambley hill (O'D.); or Hill of the boundary (Joyce).
Carrickatane (carraig a t-siain)—Rock of the foxglove.
Castlemellan (caiseal Meallain)—Mellon's stone fort.
Castlewarren (caiseal Warren)—Warren's stone fort.
Cavanoreagh (cabhan na criche)—Round hill of the boundary or territory.
Cullion (cuileann)—Ground growing holly.

Carrackyne (carraig eidhinn)—Ivy rock.
Clogherny (clochernach)—Rocky land.
Donaghedy (domhach Chaeide)—St. Caeide's church.
Drain (drainn)—A great round hill.
Drumenny (druim eanaigh)—Ridge of the marsh.
Drumgauty (druim gaoithe)—Windy ridge or top.
Dullerton (dol airtean)—The stone road.
Dunnalong (dun na long)—Fortress of the ships.
Drumman (dromann)—Ridge or long hill.
Dunnamanagh (dun na manach)—Fortress of the monks.
Dunnyboe (dun na m-bo)—The cow's fort.
Doorat (dubh rath)—Black rath or dun.
Eden (eadan)—A hill face.
Fawney (fanaidh)—A slope or declivity.
Gortaclare (gort a' clair)—The field of the plain.
Gortavea (gort an bheithe)—Field of birch.
Gortmessan (gort measan)—A fruitful field.
Gortmonly (gort monglach)—Rough field.
Grange (grainseach)—A grange.
Gobnascale (gob na scaile)—Point of the shadow or shade.
Gortileck (gort na lice)—Field of the great flat stone.
Glennagoorland (gleann a ghuairlcain)—Glen of the whirlwind.
Glencosh (gleann coise)—Foot shaped glen; or land at the foot of the glen.
Gortmellan (gort Mellon)—Mellon's field.
Glengarrow (gleann garbh)—Rough glen.
Killenny (coill leana)—A wood meadow (O D.).
Killyclooney (coille cluana)—A wood meadow (O'D.).
Killycurry (coill an churraich)—Wood of the curragh or marsh.
Leitrim (liath dhruim or leath tirim)—Grey ridge or a half dry place.
Liscloon (lios cluaine)—Fort of the meadow.
Lisdivin (lios dubhan)—Little black fort (Joyce).
(lios Daimhin)—Devon's fort (O'D.).
Leat (leacht)—A monument.
Lisnaragh (lios na rath)—Enclosure in the forts.
Loughash (loch easa)—Lough of the waterfall.
Legnacoppoge (lag na g-copog)—Hollow where dock leaves grow.

Moneycannon (muine ceann fhinne)—Shrubbery of the white-faced cow.
Magheramason (machaire Masain)—Massan's plain; or (machaire maisean)—Food producing plain.
Magherareagh (machaire riabhach)—Grey plain.
Meenagh (min achadh)—Fine green field.
Moyagh (maigheach)—Level land.
Meendamph (min damh)—Meadow of the oxen; or (min dubh)—Black mountain meadow.
Rousky (rusgaidh)—Marshy or fenny land.
Sollus (solas)—A place of light.
Stonnyfaulds (toin an fhaleadh)—Enclosed bottom lands.
Stranagalwilly (srath na gallbhuaile)—Holm of the English dairy place.
Stranabrosney (srath na m-brosna)—Holm of the faggots.
Stroanbrack (sruthan breac)—Speckled rivulet.
Tirconnelly (tir Connolly)—Connolly's district.
Tamnabrady (tamhnach Brady)—Brady's green field.
Tamnabryan (tamnach bruidhne)—Green field of the castle.
Tamnaclare (tamhnach clair)—Green field of the plain.
Tamnakeery (tamnach caerthainn)—Green field of the quicken tree (Joyce); or (tamnach caorach)—Green fields of the sheep.
Tirkernaghan (tir Cearnachain)—Kernaghan's district.
Taboe (taobh bo)—Hillside of the cows; or (teach bo)—Cow's house.
Tullyard (tullach ard)—High summit.

PARISH—LECKPATRICK.

Altnageerog (alt na g-ciarog)—Hill of the clocks or black bettles.
Artigarvan (ard tiogh Garbhain)—Garvan's high house
Ballyskeagh (baile sceach)—Town of the whitethorn shrubbery.
Ballee (baile liath)—Grey town (Joyce).
    (baile Aodh)—Hugh's town (O'D.)
Ballydonaghy (baile Donaghy)—Donaghy's town.
Ballylaw (baile locha)—Town of the lough or lake; or (baile laghach)—Generous town (O'D.).
Ballymagorry (baile M'Gorry)—M'Gorry's town.

STRABANE UNION.

Craignagaple (craig na g-capall)—Rocky hill of the horses.
Cloghcor (cloch chor)—Rough stone.
Coolermoney (cul le muine)—Backside towards the shrubbery.
Desert (disert)—A wilderness or deserted place; also a hermitage.
Fyfin (faithche fhinn)—A little white plot.
Gorticrum (gort crom)—A sloping, crooked field.
Greenlaw (grinneal ath)—Gravelly ford.
Keenaghan (caonachan)—Small tract of mossy land.
Knockanbrack (cnockan breac)—Small speckled hill.
Knockinarvoer (cnocan arbhair)—Small hill of the oats
Knocknahorna (cnoc na h-eorna)—Hill of the barley.
Killynaght (coill Ui Neacht)—O'Knought's wood.
Lagavaddar (lag a mheadair)—Hollow of the mether or drinking cup.
Lagavittal (lag a mhiotail)—Hollow of the ore.
Lagnagalloglagh (lag na n-galloglach)—Hollow of the gallowglasses.
Lisdoo (lios dubh)—Black fort.
Leckpatrick (leac Padraig)—Patrick's stone.
Liscurry (lios curraigh)—Fort of the moor.
Loughneas (loch an easa)—Lough of the waterfall.
Maccrackens (baile Mhic Crackain)—M'Cracken's town
Owenreagh (abhan riabhach)—Grey river.
Pullateebee (poll a tighe buidhe)—Hollow of the yellow house.
Stranisk (srath an uisge)—Watery holm.
Tullyard (tullach ard)—A high hill.

PARISH—URNEY.

Ballylennan (baile Ui Lennon)—Lennon's town.
Ballycolman (baile Ui Colmain)—Colman's town.
Berrysfort (biorra)—Pointed hills.
Ballyfatten (baile Phaidin)—Paddy's or Paudeen's town; or (baile an photain) town of the pot.
Castlegore (caiseal gabhar)—Circular stone fort of the goats.
Carricklee (carraig liath)—Grey rocky place.
Castlesessiagh (caiseal seiseadhach)—Fort of the division.
Carrickore (carraig oir)—Golden coloured rock.

TOWNLAND NAMES OF CO. TYRONE.

Craigmonaghan (carraig Ui Monaghan)—Monaghan's rock.
Clady (claidigh)—Muddy margin of a stream or river.
Cavan ('cabhan)—A hill; also means a hollow.
Creevy (craobhach)—Bushy or wooded land.
Dartans (dartan)—A place where cows or herds graze.
Dunnygowen (dun an ghabhann)—Fort or dun of the smith.
Drumeagle (druim eaglach)—Fearful ridge.
Freughlough (fraoch loch)—Lake or lough of the heather.
Glentimon (gleann t-Siomoin)—Simon's glen.
Ganvaghan (gainmheachan)—Little sandy place.
Gortlogher (gort luachra)—Field of the rushes.
Gallany (gallanach)—Place of standing stones.
Inchenny (inis eanaigh)—River holm of the marsh.
Inniscian (inis gleann)—Strong stream of water.
Kilclean (coill claon)—Sloping wood.
Kilcroagh (coill cruaiche)—Wood of the round hill.
Learmore (ladhar mor)—The big fork; a portion of land between two streams or rivers.
Lisdoo (lios dubh)—Black fort.
Magheragar (machaire ghearr)—The short plain.
Munie (muine)—A shrubbery.
Magirr (maigh ghearr)—Short plain or level tract of land.
Pullyernan (poll Ui Iarnain)—O'Hernan's pool or hollow.
Seein—Joyce translates it sidhean—A fairy mount; also (suidhe Fhinn) Finn M'Cool's seat or resting place.
Skerryglass (sceire glas)—Grey rock.
Tullydoortans (tullach duartain)—Hillocks or knolls (O'D.). Could also mean the hillock of the heifers.
Tullywhisker (tullach eiscir)—Hill of the ridge.
Tullymoan (tulaigh mona)—The mossy hill.
Urney (urnaidhe)—A place of prayer; an oratory.

PARISH—BADONEY UPPER.

Aghalane (achadh leathan)—The wide field.
Barnes (bearnach)—A gap.
Badoney (both domhnaigh)—The tent of the church or Sunday hut.
Ballynasollus (bel-atha-na solus)—The fordmouth of the lights, or town of the fort of the lights.

Bradkeel (braghad caol)—Narrow gorge or deep glen.
Clogherny (clocharnaidh)—A stony place.
Corramore (coramor)—Big round hill, or the big bog or curragh.
Castledamph (caislean damh)—Stone fort of the oxen.
Craigatuke (craig an t-seabhaich)—Hawks rock.
Carnargan (carn argain)—Stone heap of the plunderings.
Corrick (comh rac)—A confluence or meeting of waters or roads.
Craigatuke (craig an t-seabhach)—Hawks rock.
Cruckaclady (cnoc a chladaigh)—Hill of the mire.
Carrowwaoghtragh (ceathramha uachtarach)—Upper quarter.
Corratary (cor an t-searraigh)—Round hill of the foal.
Drumnaspar (druim na sparra)—Ridge of the spars or masts, or sharp rocky edges.
Dergbrough (dearg bruach)—Red overhanging glen or border.
Eden (eadan)—A hill face.
Glenerin (gleann thirim)—Dry glen; or (chaorain) or moory glen.
Glenchiel (gleann caol)—Narrow glen.
Glenga (glen gath)—Glen of the javelins or spears.
Glenroan (gleann ruadhan)—Red (land) glen. The inhabitants say it means "gleann sroine" glen of the nose, because it is the only projection for many miles along the valley.
Glashygolgan (glaise Gealgain)—Galligan's streamlet.
Goles (gabhal)—Forked land between streams or rivers
Garvagh (garbh achadh)—Rough land.
Glencoppoagagh (gleann copagach)—Glen of the dockings.
Keadycam (ceidecam)—The crooked smooth flat topped hill.
Letterbrat (leiter brat)—Wet sloping land of the brats or mantles.
Leogloghfin (lag na gloch bhfionn)—The pool of the stones.
Landahussy (lann da Hussey)—O'Hussey's house or church.
Lisnacreight (lios na creaghta)—The fort of the deep ditch or ravine.
Learden (ladhar donn)—The brown fork or hollow.

Lislea (lios liath)—Grey fort.
Leaghs (liath)—Land having grey spots, or a stony countryside.
Meenacrane (min na gcrann)—Mountain meadow of the dark wood.
Meenagarragh (mine garbh)—Rough meadow land.
Meenagorp (min na g-corp)—Mountain flat of the corpses.
Oughtboy (ucht buidhe)—Yellow breast of a hill.
Oughtdoorish (ucht dubh ruis)—Hill breast of the dark woods.
Oughtmame (ucht maidhme)—Breast front of the mountain pass, or pass of the defeat or battle.
Oughtnamwella (ucht na maola)—Breast of the streams
Quiggy (cuig)—The fifth part or division.
Sawelabeg (sabhall beg)—Little church land. The word means barn in reference to the first church built by St. Patrick.
Strahull (srath cuill)—The hazel holm.
Tullagherin (cul a' chsorainn)—Little hill of the coarse grazing. The last letter is changed by metathesis from "m" to "n." See Joyce, Vol. 2, p. 413.
Tullynadall (tulaigh na dala)—Hill of the meeting or assembly.

PARISH—CAMUS.

Bearney Glebe (bearna)—A gap.
Calhome (caol caomh)—Beautiful narrow valley.
Camus (camus)—A winding stream or river bend.
Carrigullin (carraig chuilinn)—Rock of the holly.
Cavanlee (cabhan liath)—The grey hillside.
Dergalt (dearg alt)—Red glenside.
Dernalebe (doire shleibhe)—Oak wood of the mountain or of the wet ground.
Drumnaboy (druim buidhe)—A yellow ridge; or (druim na buaidhe)—Ridge of the victory.
Edymore (eadan mor)—The big hill face.
Elagh (aileach)—A stone house or stone fort.
Evish (eibhis)—Coarse mountain pasture.
Liskinbwee (lios cinn buidhe)—Fort of the yellow head.
Lisky (lios sceach)—Fort of the whitethorn bushes.
Stragullin (srath cuilinn)—Holm of the hollys (Joyce).
  (srath Guillinn)—Guillon's holm (O'D.).
Strabane (srath ban)—White holm or strand.

Previous Publications from Moyola Books

John Mac Closkey's Statistical Reports (1821)
(The parishes of Ballinascreen, Kilcronaghan, Desertmartin, Banagher, Dungiven and Bovera in the County of Londonderry).

Hardback — bound in blue with gold lettering — 110 pages — numbered limited edition of 150.

Published jointly by Moyola Books (ISBN. 0-9511836-0-5) and Braid Books (ISBN, 0-9510085-2-8).

£5.95 (includ. p. & p.).

Pastime Postcards — Draperstown
A set of four reproductions of old black and white photographs of Draperstown (early 1900s). These four high quality postcards are presented in an attractive wallet.

75p per set
(includ. p. & p.).